The Sum of Our Days

ALSO BY ISABEL ALLENDE

The House of the Spirits

Of Love and Shadows

Eva Luna

The Stories of Eva Luna

The Infinite Plan

Paula

Aphrodite: A Memoir of the Senses

Daughter of Fortune

Portrait in Sepia

My Invented Country

Zorro

Inés of My Soul

FOR YOUNG ADULTS

City of the Beasts

Kingdom of the Golden Dragon

Forest of the Pygmies

The Sum of Our Days

Isabel Allende

Translated From the Spanish by
Margaret Sayers Peden

HARPER LUXE

An Imprint of HarperCollins*Publishers*

HarperCollins books may be purchased for educational, business, or sales promotional use. For information please write: Special Markets Department, HarperCollins Publishers, 10 East 53rd Street, New York, NY 10022.

FIRST HARPERLUXE EDITION

HarperLuxe™ is a trademark of HarperCollins Publishers

Library of Congress Cataloging-in-Publication Data is available upon request.

ISBN: 978-0-06-156310-2

08 09 10 11 12 ID/RRD 10 9 8 7 6 5 4 3 2 1

To the members of my
small tribe who allowed me
to tell their stories.

Acknowledgments

This book could not be published without the cooperation—in some cases reluctant—of the main characters in the story. As my son says: it is not easy to have a writer in the family. So thanks to all of them for putting up with my never-ending questions and for allowing me to dig deeper and deeper into their private lives. In some cases I have disguised the names and identities of certain people to protect their privacy. My very special gratitude goes to Margaret Sayers Peden, who went back and forth with the translation into English, enduring patiently the innumerable changes that I made along the way. Also to my agents, Carmen Balcells and Gloria Gutiérrez; my attentive reader, Jorge Manzanilla; and my editors Nuria Tey, in Spain, and Terry Karten in the United States. And most important, thanks to my mother and pen pal, Panchita, for our daily correspondence. Our letters keep my memories fresh.

The Capricious Muse of Dawn

There is no lack of drama in my life, I have more than enough three-ring-circus material for writing, but even so, I always approach the seventh of January with trembling. Last night I couldn't sleep. We were shaken by a storm; the wind roared among the oaks and rattled the windows of the house, the culmination of the biblical deluge of recent weeks. Some neighborhoods in our area were flooded; the firemen were not equipped to cope with such a major disaster, and neighbors waded out into the streets in water up to their waists to save what they could from the torrents. Furniture sailed down the main streets, and bewildered pets awaited their owners atop drowned cars, while reporters in helicopters captured scenes of this California winter you would have thought was a

Louisiana hurricane. In some places traffic was blocked for a couple of days, and when at last the skies cleared and the magnitude of the damage could be seen, crews of Latino immigrants were given the task of pumping out the water and removing debris by hand. Our house, set high on a hill, takes face-on the fury of the wind, which bends the palms and from time to time tore the proudest trees out by the roots, the ones that do not bow their heads, but we escaped the flooding. Occasionally at the height of the wind, capricious waves rise up and overflow the one access road, and at those times we were trapped, observing from above the unusual spectacle of the raging bay.

I like it that winter forces us to turn inward. I live in Marin County, to the north of San Francisco, twenty minutes from the Golden Gate Bridge, among hills golden in summer and emerald in winter, on the west shore of the enormous bay. On a clear day we can see two other bridges in the distance, the hazy outlines of the ports of Oakland and San Francisco, the slow-moving cargo ships, hundreds of sailboats, and gulls like white handkerchiefs. In May we begin to see a few intrepid adventurers hanging from multicolored comets gliding swiftly across the water, disturbing the quiet of Asian grandfathers who spend their afternoons fishing from the rocks. From the Pacific one does not see the narrow access to the bay, which greets the dawn wrapped in fog, and the sailors of yesteryear passed on by, never imagining the splendor hidden a little farther in. Now that entrance is crowned by the elegant Golden Gate Bridge, with its proud red towers. Water, sky, hills, and woods; that is my landscape.

It wasn't the end-of-the-world windstorm or the machine-gun hail on the roof tiles that kept me awake last night, it was the anxiety of knowing that with the light of day it would be the eighth of January. For twenty-five years, I have begun a book on that date, more from superstition than discipline. I'm afraid that if I begin on any other day the book will be a failure and that if I let an eighth of January go by without writing, I'll not be able to start for the rest of the year. January arrives after a few months without writing, months in which I've lived turned outward, in the uproar of the world, traveling, promoting books, giving lectures, surrounded by people, talking too much. Noise and more noise. Most of all I fear going deaf, not being able to hear the silence. Without silence, I'm done for. Last night I got up several times to wander through the house, using a variety of excuses, wrapped in Willie's old cashmere sweater, so worn it's become my second skin, with successive cups of hot chocolate in my hands, thinking and thinking about what I was going to write within a few hours, until the cold forced me back to bed, where Willie, bless him, lay snoring. Pressed against his naked back, I tucked my icy feet between his long, firm legs, breathing in the surprising scent of a young man that hasn't changed in all these years. He never wakes up when I press against him, only when I move away. He is used to my body, my insomnia, and my nightmares. And the same is true with Olivia, who sleeps on a bench at the foot of our bed. She never stirs. Nothing interrupts that silly dog's sleep, not the mice that sometimes creep out of their holes, or the funk the skunks emit as they make love, or the wandering souls murmuring in the darkness. If a madman armed with a hatchet should attack us, Olivia would be the last to know. When she came to us she was a wretched little beast the Humane Society had picked up from the dump. She had a broken leg and several broken ribs. For a month she hid among my shoes in the closet, shivering, but little by little she recovered from her previous ill treatment and emerged with her ears drooping and her tail between her legs. We knew then that she would never be a guard dog; she sleeps like a log.

By daybreak, finally, the wrath of the storm had ceased, but it was still raining. With the first light at the window, I showered and got dressed, while Willie, wrapped in his jaded sheik dressing gown, went to the kitchen. The smell of freshly ground coffee enveloped me like a caress. Aromatherapy. These everyday routines unite us more than the clamor of passion; when we're apart it is this silent dance we miss most. We each need to feel that the other one is near, always there in that intangible space that is ours alone. A cold dawn, coffee and toast, time to write, a dog that wags her tail, and my lover. Life could be no better. Willie gave me a good-bye hug, for I was leaving on a long journey. "Good luck," he whispered, as he does every year on this day, and I took my coat and umbrella, went down six steps, skirted the swimming pool, walked through fifty feet of garden, and reached the casita where I write, my study, my *cuchitril*. And here I am now.

I had barely lit a candle—one always illuminates my writing—when Carmen Balcells, my agent, called me from Santa Fe, a tiny town of crazed goats near Barcelona, where she was born. She intends to spend her mature years there in peace, but as she has energy to burn, she is buying the village house by house.

"Read me the first sentence," demanded this larger-than-life mother figure.

I reminded her once more of the nine-hour difference in time between California and Spain. No first sentence yet. No nothing.

"Write a memoir, Isabel."

"I already wrote one, don't you remember?"

"That was thirteen years ago."

"My family doesn't like to see itself exposed, Carmen."

"Don't worry about anything. Send me a two- or three-hundred-page letter and I'll take care of the rest. If it comes down to choosing between telling a story and offending relatives, any professional writer chooses the former."

"Are you sure?"

"Absolutely."

PART ONE

Darkest Waters

In the second week of December, 1992, almost as soon as the rain let up, we went as a family to scatter your ashes, Paula, following the instructions you had left in a letter written long before you fell ill. As soon as we advised them of your death, your husband, Ernesto, came from New Jersey, and your father from Chile. They were able to tell you good-bye where you lay wrapped in a white sheet waiting to be taken to the crematory. Afterward, we met in a church to hear mass and weep together. Your father was pressed to return to Chile, but he waited until the weather cleared, and two days later, when finally a timid ray of sun peered out, the whole family, in three cars, drove to a nearby forest. Your father went in the lead, guiding us. He isn't familiar with this region but he had spent the previous two days looking for the best site, one that you would have chosen. There are many places to choose from, nature is prodigal here, but by one of those coincidences that now are habitual in anything related to you, he led us directly to the forest where I often went to walk to ease my rage and pain while you were sick, the same one where Willie had taken me for a picnic shortly after we met, the same one where you and Ernesto liked to walk hand in hand when you came to visit us in California. Your father drove into the park, followed the road a little way, parked the car, and signaled us to follow him. He took us to the exact spot that I would have chosen, because I had been there many times to pray for you: a stream surrounded with tall redwoods whose tops formed the dome of a green cathedral. There was a fine, light mist that blurred the contours of reality: the light barely penetrated the trees, but the branches shone, winter wet. An intense aroma of humus and dill rose from the earth. We stopped at the edge of a pond formed by rocks and fallen tree trunks. Ernesto, serious, haggard, but now without tears because he had spilled them all, held the clay urn containing your ashes. I had saved a few in a little porcelain box to keep forever on my altar. Your brother, Nico, had Alejandro in his arms, and your sister-in-law, Celia, held Andrea, still a baby, wrapped in shawls and clamped to her breast. I carried a bouquet of roses, which I tossed, one by one, into the water. Then all of us, including Alejandro, who was three, took a handful of ashes from the urn and dropped them onto the water. Some floated briefly among the roses, but most sank to the bottom, like fine white sand.

"What is this?" Alejandro asked.

"Your aunt Paula," my mother told him, sobbing.

"It doesn't look like her," he commented, confused.

I will begin by telling you what has happened since 1993, when you left us, and will limit myself to the family, which is what interests you. I'll have to omit two of Willie's sons: Lindsay, whom I barely know—I've seen him only a dozen times and we've never exchanged more than the essential courteous greetings—and Scott, because he doesn't want to appear in these pages. You were very fond of that thin, solitary boy with thick eyeglasses and disheveled hair. Now he is a man of twenty-eight; he looks like Willie and his name is Harleigh. He chose the name Scott when he was five; he liked it and used it a long time, but during his teens he reclaimed the one given him.

The first person who comes to my mind and heart is Jennifer, Willie's only daughter, who at the beginning of that year had just escaped for the third time from a hospital where she had gone to find rest for her bones because of yet another infection, among the many she had suffered in her short life. The police had not given any indication that they were going to look for her; they had too many cases like hers, and this time Willie's contacts with the law didn't help at all. The physician, a tall, discreet Filipino who by dint of perseverance had saved her when she arrived at the hospital with a raging fever, and who by now knew her because he had attended her on two previous occasions, explained to Willie that he had to find his daughter soon or she would die. With massive doses of antibiotics for several weeks, he might be able to save her, he said, but we had to prevent a relapse, for that would be fatal. We were in the emergency room—yellow walls, plastic chairs, and posters of mammograms and tests for AIDS— which was filled with patients awaiting their turn to be treated. The doctor took off his round, metal-framed glasses, cleaned them with a tissue, and guardedly answered our questions. He had no sympathy for Willie or for me; he perhaps mistook me for Jennifer's mother. In his eyes we were guilty; we had neglected her, and now when it was too late, we had showed up acting distressed. He avoided going into details—patient information was confidential—but Willie could deduce that in addition to multiple infections and bones turned to splinters, his daughter's heart was on the verge of giving out. For nine years Jennifer had persisted in jousting with death.

We had been going to see her in the hospital for several weeks. Her wrists were tied down so that in the delirium of fever she couldn't tear out the intravenous tubes. She was addicted to nearly every known drug, from tobacco to heroin. I don't know how her body had endured so much abuse. Since they couldn't find a healthy vein in which to inject medications, they had implanted a port in an artery in her chest. At the end of a week they had moved Jennifer from the intensive care unit to a three-bed room she shared with other patients, where she was no longer restrained, and where she was not watched as closely as she had been before. I started visiting every day, bringing things she had asked for: perfumes, nightgowns, music, but it all disappeared. I supposed that her buddies were coming at strange hours to furnish her drugs, which, since she had no money, she paid for with my gifts. As part of her treatment, she was given methadone to help her through withdrawal, but in addition to that she was using any drug her providers could smuggle to her—and which she injected straight into the port. Sometimes it was I who bathed her. Her ankles and feet were swollen, her body covered with bruises, marks from infected needles, and a scar worthy of a pirate on her back. "A knife," was her laconic explanation.

Willie's daughter was a blonde with large blue eyes like her father's, but few photographs have survived from the past and no one remembered her as she had been: the best student in her class, obedient, and well groomed. She seemed ethereal. I met her in 1988, shortly after moving to California to live with Willie, a time when she was still beautiful, although she already had an evasive look and that deceptive fog that encircled her like a dark halo. My head was spinning with my newly inaugurated love affair with Willie, and I was not overly surprised when one winter Sunday he took me to a jail on the east side of San Francisco Bay. We waited a long time on an inhospitable patio, standing in line with other visitors, most of them blacks and Latinos, until the gates were opened and we were allowed to enter a gloomy building. They separated the few men from the many women and children. I don't know what Willie's experience was, but a uniformed matron confiscated my handbag, pushed me behind a curtain, and put her hands where no one had dared, more roughly than was necessary, perhaps because my accent made her suspicious. Luckily a Salvadoran peasant woman, a visitor like me, had warned me in the line not to make a fuss, because that would make things worse. Finally

Willie and I met in a trailer set up for visiting the prisoners, a long narrow space divided by hen-coop wire; Jennifer sat behind that. She had been in jail, without drugs and well nourished, for two months. She looked like a schoolgirl in Sunday clothes, in contrast with the rough appearance of the other prisoners. She greeted her father with unbearable sadness. In the years that followed I came to realize that she always cried when she was with Willie, whether from shame or rancor I don't know. Willie introduced me briefly as "a friend," although we had been living together for some time, and stood before the wire with crossed arms and eyes cast on the floor. I watched them from a short distance away, listening to bits of their dialogue through the murmurs of other voices.

"What's it for this time?"

"You already know that, why do you ask me. Get me out of here, Dad."

"I can't."

"But you're a lawyer, aren't you."

"I warned you the last time that I wouldn't help you again. If you have chosen this life, you have to pay the consequences."

Jennifer wiped away tears with her sleeve, but they kept running down her cheeks as she asked about her brother and her mother. Soon they said good-bye, and she was led out by the same uniformed woman who had taken my handbag. At that time she still had a shred of innocence, but six years later, when she escaped from the care of the Filipino doctor in the hospital, there was nothing left of the girl I had met in that prison. At twenty-six, she looked sixty.

When we left the jail it was raining and we ran, soaking wet, the two blocks to the parking garage where we had left the car. I asked Willie why he treated his daughter so coldly, why he didn't place her in a rehabilitation program instead of leaving her behind bars.

"She's safer there," he replied.

"Can't you do anything? She has to have some treatment!"

"It's pointless, she has never wanted to accept help, and I can't force her, she's of age."

"If she were my daughter I would move heaven and earth to save her."

"She isn't your daughter," he told me with a kind of mute resentment.

At the time, a young Christian was hanging around Jennifer, one of those alcoholics redeemed by the message of Jesus, who have the same fervor for religion they once had devoted to the bottle. We saw him occasionally at the prison on visiting day, always with Bible in hand and wearing the beatific smile of God's chosen. He greeted us with the compassion reserved for those who live in the darkness of error, a tone that drove Willie to frenzy but that had the desired effect on me: it made me feel guilty. It takes very little to make me feel guilty. Sometimes he took me aside to talk to me. While he was quoting the New Testament—"*Jesus said to the sinning woman, Let he who is without sin cast the first stone*"—I observed his bad teeth with fascination and tried to protect myself from his saliva spray. I have no idea how old he was, but when he wasn't talking he seemed very young—maybe because of his freckles and the way he reminded me of a cricket—but that impression disappeared the minute he began preaching, all exorbitant gestures and strident voice. At first he tried to draw Jennifer into the ranks of the just by using the logic of his faith, but she was immune; then he opted for modest gifts, which had better results. For a handful of cigarettes she would tolerate for a while his reading passages from the Gospel.

When Jennifer got out of jail, he was waiting at the gate of the prison, dressed in a clean shirt and reeking of cologne. He often called us late at night to give us news of his protégée, and to threaten Willie, telling him he must repent of his sins and accept the Lord in his heart; then he could receive the baptism of the elect and rejoin his daughter in the shelter of divine love. He did not know whom he was dealing with: Willie is the son of a bizarre preacher; he grew up in a tent where his father, with a fat, tame snake rolled about his waist, imposed on believers his invented religion, which was why when he sniffed even a hint of a sermon, he split. The evangelical was obsessed with Jennifer, drawn to her like a moth to the flame. He was torn between his mystic fervor and his carnal passion, between saving the soul of this Mary Magdalene and taking his pleasure of her somewhat battered, but still exciting, body, something he confessed to us with such candor that we could not make fun of him. "I shall not fall into the delirium of concupiscence; no, I shall wed her," he assured us with that strange vocabulary he had, and immediately treated us to a lecture on chastity in matrimony that left us speechless. "This guy is either a complete idiot or he's gay," was Willie's comment, but he nevertheless clung to the idea of that marriage because that good-intentioned wretch might rescue his daughter. However, when her suitor, on his knees, set forth his plan to Jennifer, her response was a burst of laughter. That preacher was killed, brutally beaten to death, in a bar in the port, where he had gone one night to look for Jennifer and to spread the gentle message of Jesus among sailors and stevedores who were not in the humor to be led to Christ. We were never again awakened at midnight to listen to his messianic sermons.

Jennifer had spent her childhood hiding in corners, invisible, while her brother Lindsay, two years older than she, monopolized the attention of his parents, who could not control him. She was a mysterious and well-behaved little girl with a sense of humor too sophisticated for her years. She laughed at herself with clear, contagious giggles. No one suspected that she was climbing out a window at night until she was arrested in one of the most sordid neighborhoods in San Francisco many miles from her house, an area where the police are afraid to venture at night. She was fifteen. Her parents had been divorced for several years; both were occupied in their own affairs and perhaps had not gauged the gravity of Jennifer's problem. Willie was hard-pressed to recognize the heavily made-up girl shivering in a cell in the police station, unable to stand up or speak a word. Hours later, safe in her bed and with her mind a little clearer, Jennifer promised her father that she was going to do better and would never do anything that foolish again. He believed her. All kids stumble and fall; he too had had problems with the law when he was a boy. That had been in Los Angeles, when he was thirteen, and his offenses were stealing ice cream and smoking marijuana with the Mexican kids in the barrio. At fourteen he had realized that if he didn't straighten up right away he would be in trouble all his life, because he had no one who could help him, so he kept his distance from gangs and made up his mind to finish school, work his way through the university, and become a lawyer.

After Jennifer fled the hospital and the efforts of the Filipino doctor, she survived because she was very strong, despite her seeming fragility, and we heard nothing of her for a while. Then one winter day we heard a vague rumor that she was pregnant, but we rejected it as being impossible. She herself had told us she couldn't have children, her body had suffered too much abuse. Three months later she came to Willie's office to ask for money, something she rarely did since she preferred to make her own way—in that case she didn't have to offer explanations. Her eyes darted around desperately, looking for something she couldn't find; her hands were trembling but her voice was strong.

"I'm pregnant," she announced.

"That can't be!" Willie exclaimed.

"That's what I thought, but look . . ." She unbuttoned the man's shirt that covered her to the knees and showed him a protuberance the size of a grapefruit.

"It will be a girl and she will be born this summer. I will call her Sabrina. I've always liked that name."

Every Life a Melodrama

I spent nearly all of 1993 off to myself writing to you, Paula, crying and remembering, but I wasn't able to avoid a long book tour through several American cities promoting *The Infinite Plan*, a novel inspired by Willie's life. It had just been published in English though it had been written two years before and had already appeared in several European languages. I stole the title from Willie's father, whose wandering religion was called the Infinite Plan. Willie had ordered the book as a gift for all his friends—by my calculations he had bought out the complete first printing. He was so proud that I had to remind him that it wasn't his biography, it was fiction. "My life is a novel," was his answer. Every life can be told as a novel; each of us is the protagonist of his own legend. At this moment, writing these pages,

I have doubts. Did things really happen as I remember them and as I am telling them? Even with my mother's invaluable correspondence, which preserves a daily, and more or less truthful, version of essential as well as trivial events, these pages are subjective. Willie told me that *The Infinite Plan* was a map of his course through life, and added that it was a shame that the actor Paul Newman was a little old to play the part of the protagonist, in case it was made into a film. "You must have noticed that Paul Newman looks like me," he pointed out with his usual modesty. I hadn't noticed, but I didn't know Willie when he was young, when surely they were as alike as two peas in a pod.

The publication of the book in English came at a bad time for me; I didn't want to see anyone, and the idea of a book tour frightened me. I was still sick with grief, obsessed by what I might have done, but hadn't, to save you. Why hadn't I recognized the incompetence of the medical staff in that hospital in Madrid? Why hadn't I immediately taken you out of there and brought you back home with me to California? Why? Why? I closed myself in the room where you lived your last days, but not even in that sacred place did I find peace. Many years would go by before you became a gentle, constant friend. In those days I felt your absence as a sharp pain that at times brought me to my knees.

I was also worried about Nico because we had just learned that he, too, had porphyria. "Paula didn't die of the condition, but from medical negligence," he insisted to calm me, but he was uneasy, not so much for himself as for his two children, and the third that was on the way; that ominous heritage might have been passed on to them, we would know that when the children were old enough to undergo the tests. Three months after your death, Celia announced that she was expecting another child, something I had already suspected because of the somnambulist's purple circles beneath her eyes and because I had dreamed it, just as I had dreamed Alejandro and Andrea before they moved in their mother's womb. Three little ones in five years was ill considered, given that she and Nico did not have steady jobs and also that their student visas were about to expire, but we celebrated nonetheless. "Don't worry. Every child comes into the world with a loaf of bread under its arm," was my mother's comment when she heard. And so it was. That same week we started on the paperwork to obtain residency visas for Nico and his family, thanks to the fact that after five years of waiting I was finally an American citizen and could sponsor them.

Willie and I had met in 1987, three months before you met Ernesto. Someone told you that I had left your father for him, but I promise you, that wasn't how it was. Your father and I were together twenty-nine years; we met when I was fifteen and he was soon to be twenty. When we decided to get a divorce I hadn't the slightest suspicion that a few months later I would stumble upon Willie. We were brought together by literature. He had read my second novel and was curious to meet me when I sped like a comet across northern California. He was more than a little disappointed when he saw me because I am not at all the kind of woman he prefers, but he put up a good front and today he assures me that when he saw me he immediately felt a "spiritual connection." I don't know what that would be. As for me, I had to act fast, because I was leaping from city to city on a crazed tour. I called you to ask your advice and you answered, screaming with laughter, why on earth was I asking you if I'd already made the decision to throw myself headfirst into the adventure. I told Nico, and he exclaimed with horror: "At your age, Mamá!" I was forty-five, which to him seemed the threshold of the tomb. That was my clue that I had no time to waste, I had to get down to serious business. My urgency erased Willie's justifiable caution. I won't repeat here what you already know and I have told you many times. According to Willie I have fifty versions of how our love began . . . and all of them are true. In summary, I will say only that a few days later I abandoned my former life and, uninvited, turned up at the door of that man with whom I was so infatuated. Nico says that I "abandoned my children," but you were studying in Virginia and he was already twenty-one years old, a fine young man who was past needing to be coddled by his mother. Once Willie had recovered from his shock at seeing me on his threshold, with my suitcase, we began our lives together with enthusiasm, despite the cultural differences that separated us and the problems of his children, whom neither he nor I knew how to deal with. It seemed to me that Willie's life and family were like a bad comedy in which nothing went as it should. How many times did I call you to ask your council? I think every day. And you always gave me the same answer: "What is the most generous thing you can do in this case, Mamá?" Willie and I were married eight months later. And not at his initiative, but mine. When I realized that the passion of those first moments was turning into love, and that probably I would be staying in California, I decided to bring my children to the United States. If I wanted to be reunited with you and Nico I would have to be a citizen, so I had no choice but to swallow my pride and suggest the idea of marriage to Willie. His reaction was not the explosive joy I perhaps had dared hope for, rather more like terror; several failed love affairs had cooled the coals of romance in his heart, but in the end I twisted his arm. Well, in fact it wasn't that difficult; I gave him until noon the next day to decide and began to pack my suitcase. Fifteen minutes before the time ran out, Willie agreed to marry me, although he never understood my stubborn insistence on living near Nico and you because in the United States children abandon their family home when they finish school and return only for a visit at Christmas or Thanksgiving. Americans are shocked by the Chilean custom of living as a clan all their lives.

"Don't make me choose between my children and you!" I warned on that occasion.

"I wouldn't think of it. But are you sure that they want to live near you?" he asked.

"A mother always has the right to gather her children around her."

We were married by a man who had obtained his license through the mail by paying twenty-five dollars because though Willie was a lawyer, he couldn't find any of his judge friends to do it. That made me apprehensive. It was the hottest day in the history of Marin County. The ceremony took place in an Italian restaurant that didn't have air-conditioning; the cake melted down to nothing, the woman who was playing the harp fainted, and the guests, streaming sweat, started taking off their clothes. The men ended up without shirts or shoes and we women with no stockings or underwear. I didn't know a soul except your brother and you, my mother, and my American editor, all of whom had come a great distance to be with me. I've always suspected that the marriage was not completely legal, and hope that some day we'll have the enterprise to be married properly.

I don't want to give the impression that I married for convenience alone, since I felt for Willie the heroic lust that tends to make women of my breed lose their heads—that's how you felt about Ernesto—but at the age I was when we met, there was no need to marry were it not for the matter of the visas. In other circumstances we would have lived without the sanction of marriage, as Willie would doubtlessly have preferred, but I had no thought of renouncing my family, no matter how much my reluctant lover resembled Paul Newman. I had left Chile with you and Nico during the military dictatorship of the '70s; together we had found refuge in Venezuela until the end of the '80s; and with the two of you in the '90s, I intended to become a United States immigrant. There was no question in my mind that your brother and you would be much better off with me in California than scattered across the world, but I had failed to take legal delays into account. Five years went by, which were like five centuries, and in the meantime you both had married, Nico to Celia in Venezuela, and you to Ernesto in Spain, but that didn't seem a serious impediment to my plan. After some time went by, I had succeeded in installing Nico and his family two blocks from where we lived, and if death had not snatched you away long before your time, you, too, would have lived nearby.

I left on a book tour, crisscrossing the United States to promote my novel and give the readings and lectures that had been postponed the year before when I was unable to leave your side. Did you feel my presence, Paula? I've asked myself that many times. What were you dreaming that long night in 1992? I'm sure you dreamed because your eyes moved behind your eyelids and at times you awoke frightened. Being in a coma must be like being trapped in the dense fog of a nightmare. According to physicians, you weren't aware of anything, but that was always hard for me to believe.

On the trip I carried a bag of pills for sleeping, pills for imagined pains, pills for drying my tears, and pills for my fear of loneliness. Willie couldn't go with me because he had to work; his office never closed, not even Sundays, and there was always an assortment of miscreants seeking miracles in his waiting room and a hundred cases on his desk. Just at that time, he was deeply involved with the tragedy of a Mexican immigrant who had died when he fell from the fifth floor of a building under construction in San Francisco. His name was Jovito Pacheco, and he was twenty-nine years old. Officially, he didn't exist. The construction company had washed its hands of him, the man's name was not on its payroll. The subcontractor had no insurance, and he, too, had "never seen" Pacheco. Actually, he had recruited him a few days before from his truck, along with twenty other illegals like him, and had driven him to the work site. Jovito Pacheco was a campesino and had never climbed a scaffold, but he had strong shoulders and a strong desire to work. No one told him that he should put on a safety harness. "I'll sue half the world if I have to, but I'm going to get some compensation for that poor family!" I heard Willie say a thousand times. Apparently it wasn't an easy case. He had a faded photograph of the Pacheco family in his office: father, mother, grandmother, three small children, and a babe in arms, all dressed in their Sunday finest and lined up in the bright sun of a dusty plaza in Mexico. The only one wearing shoes was Jovito Pacheco, a dark-skinned Indian with a proud smile and a straw hat in his hand.

On that tour I wore black from head to foot, using the pretext that black is an elegant color; I did not want to admit, even to myself, that I was wearing mourning. "You look like a Chilean widow," Willie told me, and gave me a fire-truck-red scarf. I don't remember what cities I went to, whom I met, or what I did; none of it mattered anyway, except for my meeting in New York with Ernesto. Your husband was very moved when I told him I was writing a memoir about you. We wept together, and the sum of our grief was unleashed in the form of a hailstorm. "It often hails in winter," Nico commented when I told him over the phone.

I spent several weeks far from my loved ones, moving as if I were hypnotized. At night I would fall into strange beds, anesthetized with sleeping pills, and in the mornings I shook off my bad dreams with black coffee. I spoke by phone with everyone in California and sent my mother letters by fax that faded with time because they were printed with an ink sensitive to light. Much of what happened during that time was lost. I'm sure it's better so. I counted the hours until I could go back home and hide from the world. I wanted to sleep next to Willie, play with my grandchildren, and console myself making necklaces in my friend Tabra's workshop.

I found that in her pregnancy Celia was losing weight instead of gaining it, that my grandson Alejandro was going to day care with a backpack like a big boy, and that Andrea needed an operation on her eyes. My granddaughter was very small, with a head of curly golden hair and eyes that were completely crossed; her left eye simply went its own way. She was very quiet and didn't romp around; she seemed always to be planning something, and as she sucked a finger she clung to a cotton diaper—her *tuto*—which she seldom let out of her hands. You never liked children, Paula. Once when you came to visit and you had to change Alejandro's diaper, you confessed to me that the more you were with the baby, the less desire you had to be a mother. You never knew Andrea, but the night you died she was sleeping, beside her brother, at the foot of your bed.

An Old Soul Comes to Visit

In May Willie called me in New York to tell me that, defying the predictions of science and the law of probability, Jennifer had given birth to a little girl. A double dose of narcotics had precipitated the birth, and Sabrina had been born two months before term. Someone had called an ambulance, which took Jennifer to the nearest emergency room, a private Catholic hospital where they had never seen anyone in that state of intoxication. That saved Sabrina, because had she been born in the public hospital in the poor section of Oakland where Jennifer lived, she would have been just one more of the hundreds of babies born only to die, condemned by drugs in the maternal womb. No one would have noticed her, and the tiny infant would have been lost in the cracks of the overloaded social medical system.

Instead, she fell into the skilled hands of the emergency room physician who received her when she was spit out into the world, and who in the process became the first person to be seduced by Sabrina's hypnotic gaze. "This child has little chance to live," was his diagnosis when he examined her, but he was entangled in the web of her dark eyes and that evening did not go home at the end of his shift. By then a pediatrician had arrived, and the two of them stayed part of the night, keeping watch over the incubator and attempting to figure out how to detox this newborn without harming her more than she already had been, as well as how to feed her, since she couldn't yet swallow. They had no time to worry about the mother; she had fled the hospital as soon as she could get out of the bed.

Jennifer had been struck by a pain that threatened to split her apart, and she didn't remember much of what had happened, only the terrifying shriek of the ambulance's siren, a long corridor with bright lights, and faces shouting orders. She thought she had given birth to a girl, but she couldn't stay to confirm it. They had left her resting in a room, but very soon she had felt symptoms of withdrawal and had begun to shake with nausea; she was bathed in sweat, and her nerves were live electric wires. She had dressed however she could and escaped through a service door. A couple of days later, somewhat recovered from the delivery and tranquilized by drugs, she thought of the infant she had left in the clinic and went back to look for her. But Sabrina was no longer hers. The Child Protective Services had intervened and put a monitor on Sabrina's arm that would activate an alarm if anyone tried to take her from the room.

I interrupted my tour in New York and returned on the first available flight to California. Willie picked me up at the airport, drove me directly to the hospital, and along the way explained that his granddaughter was very ill. Jennifer, lost in her own purgatory, could not take care of herself, say nothing of take charge of her daughter. She lived with a man twice her age who had been arrested more than once. "I'm sure he's exploiting Jennifer and getting her drugs," was my first thought, but Willie, who is much nobler than I, was grateful that the man at least provided a roof over Jennifer's head.

We ran down the corridors of the hospital to the nursery for the premature babies. The nurse already knew Willie, and took us to a little cradle back in one corner. I first took Sabrina in my arms one warm day in May; she was wrapped in a cotton blanket, like a little package. I opened the bundle fold by fold, and in its depths found a little curled-up snail in a diaper that enveloped her from her ankles to her neck. Two tiny wrinkled feet, arms like toothpicks, and a perfect head covered in a knit cap stuck out of the diaper; she had fine features and large, dark, almond-shaped eyes that stared at me with the determination of a warrior's. She weighed nothing at all. Her skin was dry and smelled of medications; she was soft, pure foam. "She was born with her eyes open," the nurse told us. Sabrina and I observed each other for a long time, getting acquainted. They say that at that age babies are nearly blind, but she had the same intense expression that characterizes her today. I held out a finger to stroke her cheek and her tiny fist grabbed on to it. I could feel her shivering, and I wrapped her back in her little blanket and held her tight against my breast.

"How are you related to the baby?" asked a young woman who had introduced herself as the hospital's pediatrician.

"He's her grandfather," I replied, nodding toward my husband, who was over by the door, timid, or too emotional to speak.

"Our tests reveal the presence of various toxic substances in the baby's system. She is also premature; by my calculations she must be barely thirty-two weeks; she weighs three and a half pounds and her digestive system is not totally developed."

"Shouldn't she be in an incubator?" Willie inquired.

"We took her out of the incubator today because her respiration is normal. But don't get your hopes up. I'm afraid the prognosis is not very good—"

"She's going to live!" the nurse interrupted emphatically as she took Sabrina from me. She was a majestic African-American woman with a tower of tiny braids atop her head and plump arms into which Sabrina promptly disappeared.

"Odilia, please!" exclaimed the pediatrician, incredulous at this totally unprofessional eruption.

"That's all right, Doctor, we understand," I told her with a weary sigh.

I hadn't had time to change the dress I'd been wearing for weeks on my tour. I had visited fifteen cities in twenty-one days, carrying a small tote that contained the essentials, which in my experience is very little. I would take a plane at the first light of day, reach the appropriate city, where an escort—nearly always a woman as exhausted as I—was waiting to take me to appointments with the press. I would eat a sandwich at noon, have a couple of interviews more, and go to the hotel to shower before the night's program, in which I faced the public with swollen feet and a forced smile and read a few pages of my novel in English. I carried a framed photograph of you so you would be with me in the hotels. I wanted to remember you that way, with your splendid smile, your long hair, and your green blouse, but when I thought about you the images that assaulted me were other: your stiff body, your empty eyes, your absolute silence. In those publicity marathons, which would pulverize the bones of an Amazon, I traveled out of body, as if on an astral journey, and fulfilled the stages of the tour with a heavy rock in my chest, confident that my escorts would lead me by the hand during the day, accompany me to that night's reading, and leave me at the airport the following dawn. During the long hours of the flight from New York to San Francisco, I had time to think about Sabrina, but I never imagined the way that granddaughter would change lives.

"She has a very old soul," said Odilia, the nurse, after the pediatrician left. "I've seen a lot of newborns in the twenty-two years I've worked here, but one like Sabrina . . . never. She takes in everything. I stay with her even after my shift is over, and I came Sunday to see her, because I can't get her out of my head."

"Do you think she's going to die?" I asked in a choked voice.

"That's what the staff say. You heard the doctor. But I know she will live. She's come to stay; she has good karma."

Karma. Again karma. How many times have I heard that term in California? The idea of karma drives me up the wall. To believe in destiny is limiting enough, but karma is much worse; it goes back through a thousand previous lifetimes, and sometimes you have to carry the misdeeds of your ancestors. Destiny can be changed, but to clear your karma takes a lifetime, and even that may not be enough. But that wasn't the moment to discuss philosophy with Odilia. I felt an infinite tenderness for the baby and gratitude for the nurse who felt real affection for her. I buried my face in the diaper that enveloped Sabrina, happy that she was in the world.

Willie and I left the nursery holding each other up. We went down several identical corridors looking for the exit, until we came to an elevator. Its mirror returned our images. It seemed to me that Willie had aged a century. His shoulders, always so arrogant, now slumped in defeat. I noticed the wrinkles around his eyes, the line of his jaw, less bold than before, and how at some point his little remaining hair had turned white. How quickly the days go by. I hadn't seen the changes in his body and I didn't see him as he was but as I remembered him. To me he was still the man I had fallen in love with at first sight six years before: handsome, athletic, wearing a dark suit that fit him a little snugly, as if his shoulders were challenging the seams. I liked his spontaneous laughter, his confident attitude, his elegant hands. He inhaled all the air around him, occupied all the space. One could see that he had lived and suffered, but he seemed invulnerable. And me? What had he seen in me when we met? How much I had changed in those six years, especially these last months? I had also been seeing myself through the same charitable filter of habit, never noticing the inevitable physical decline, the less firm breasts, the thicker waist, the sadder eyes. The mirror in the elevator revealed to me how exhausted we both were, something more profound than weariness from my travel and his work. Buddhists say that life is a river, that we are carried on a raft to our final destination. The river has its current, rapids, sandbars, whirlpools, and other obstacles that we can't control, but we are given a pair of oars to guide our craft. The quality of the voyage depends upon our skill, but we cannot alter the course because the river always empties into death. Sometimes we have no choice but to give ourselves to the current, but that wasn't the case here. I took a deep breath, stretched to my full, albeit meager, height, and slapped my husband on the back.

"Stand up straight, Willie, we have to row."

He looked at me with that confused expression he tends to have when he thinks my English is deserting me.

A Nest for Sabrina

I never doubted that Willie and I would take charge of Sabrina: if the parents can't do it, it becomes the responsibility of the grandparents; it's a law of nature. However, I soon discovered that it would not be that simple. It wasn't just taking a basket to the hospital to pick up the baby when they released her in a month or two. There were matters to be taken care of. The judge had already determined that she could not be handed over to Jennifer, but the man she lived with was still in the picture. I didn't believe that he was the father because Sabrina didn't have his African features—though I was assured that she was not purely Caucasian and that her skin would darken over the course of the weeks. Willie asked for a blood test, and although the man refused to take one, Jennifer had confirmed that he was the father, and that was all that was needed legally. From Chile, my mother advised us that it would be insane for us to adopt Sabrina, that Willie and I were worn too thin for a task of such magnitude. Willie had enough problems with his children and his office, and I had no break in my writing and traveling.

"That baby will have to be cared for day and night. How do you plan to do that?" she asked.

"The same way I cared for Paula," I pronounced.

Nico and Celia came to talk with us. Your brother, slim as a birch and still with the face of a runny-nosed kid, had a child in each arm. It was obvious from her belly that Celia was six months pregnant; she looked tired and her skin was sallow. Once again, I was amazed when I looked at Nico, who inherited nothing from me; he is a head and a half taller than I am, composed and rational, he has elegant manners, and is blessed with a gentle sense of irony. His intellect is pristinely clear, focused not only on mathematics and science, which are his passions, but also on any human activity. I am constantly surprised by what he knows, by his opinions. He finds solutions for all kinds of problems, from a complex computer program to another, no less complex mechanism for hanging a bicycle from the ceiling with no fuss. He can fix almost any object of practical use, and does it with such care that it comes out better than it was originally. I have never seen him lose control. He has three basic rules that he applies in his relationships: it isn't personal, everyone is responsible for his or her own feelings, life isn't fair. Where did he learn that? From the Mafia, I suppose. Don Corleone. I have tried in vain to follow his path of wisdom but . . . for me everything is personal, I do feel responsible for the feelings of other people, even those I scarcely know, and I have for more than sixty years been frustrated because I can't accept that life is unfair.

You had very little time to know your sister-in-law well, and I suspect that you weren't overly fond of her since you were rather stern. I was a little afraid of you myself, Paula, I can tell you that now: your judgments tended to be concise and irrevocable. Besides, Celia raised people's dander on purpose, it was as if she took great pains to shock everyone. Let me remind you of one conversation at the table.

"I think they ought to ship all the queers to an island and make them stay there. It's their fault that we have the AIDS epidemic," said Celia.

"How can you say something like that!" you exclaimed, horrified.

"Why do we have to pay for those people's problems?"

"What island?" Willie asked, to be annoying.

"I don't know. The Farallons, for example."

"The Farallons are very small."

"Any island! A gay island where they can take it in the ass until they die!"

"And what would they eat?"

"Let them plant their vegetables and tend their chickens! Or we can use tax money to set up an airlift."

"Your English has improved a lot, Celia. Now you can articulate your bigotry to perfection," my husband commented with a broad smile.

"Thank you, Willie," she replied.

And that was how the conversation went as we sat around talking, until you left, indignant. It's true. Celia tended to express herself in rather bold fashion, at least for California, but we have to remember that for several years she was involved with the Opus Dei, and that she came from Venezuela, where no one's tongue is tied when it comes to saying anything they want. Celia is intelligent and contradictory; she has tremendous energy and an irreverent sense of humor that, translated into the limited English she had at that time, caused havoc. She worked as my assistant, and more than one journalist or unwarned visitor left my office put off by my daughter-in-law's jokes. But I want to tell you something you may not know,

Paula: she looked after you for months with the same tenderness she devoted to her children; she was with you in your last hours; she helped me prepare your body in the intimate rites of death; and she stayed beside you, waiting a day and a night, until Ernesto and the rest of the family that had traveled long distances arrived. We wanted you to receive them in your bed, in our house, for the final good-bye. But back to Sabrina. Nico and Celia joined us in the living room, and for once she had nothing to say; her eyes were glued on her wool socks and Franciscan monk's sandals. It was Nico who did the talking. He began with my mother's argument that Willie and I were not of an age to be taking on the care of a baby. When Sabrina was fifteen, I would be sixty-six and Willie seventy-one.

"Willie is no genius when it comes to raising children, and you, Mamá, you're trying to replace Paula with a sick little baby. Would you be strong enough to bear grief like that again if Sabrina doesn't survive? I don't think so. But we're young, and we can do it. We've already talked it over and we're prepared to adopt Sabrina," my son concluded.

For a long moment, Willie and I couldn't speak.

"But very soon you're going to have three children of your own," I managed to say finally.

"And what is one more stripe to the tiger?" Celia mumbled.

"Thank you, I really do thank you, but that would be madness. You have your own family and you need to get ahead in this country, which won't be easy. You can't be responsible for Sabrina, that's up to us."

In the meantime, behind our backs the days were going by and the cumbersome machinery of the law was following its inexorable course. The social worker in charge of the case, Rebecca, looked very young, but she had had a lot of experience. Her job was not one to be envied; she had to work with children who had suffered abuse and neglect, children who were shuttled from one institution to the next, who were adopted and then returned, children terrorized and filled with rage, children who were delinquent, or so traumatized that they would never lead a more or less normal life. Rebecca fought the bureaucracy, the institutionalized negligence, the lack of resources, the irremediable wickedness of humankind, and, especially, she fought time. There weren't enough hours to study cases, visit the children, rescue the ones in the most urgent danger, find them a temporary refuge, protect them, save them, follow their cases. The same children passed through her office again and again, their problems growing worse with the years. Nothing was resolved, only postponed. After reading the information she had before her, Rebecca decided that when Sabrina left the hospital she should be sent to a foster home that specialized in children with serious illnesses. She filled out the necessary documents, they leaped from desk to desk until they reached the proper judge, and he signed them. Sabrina's fate was sealed. When I learned that, I flew to Willie's office, pulled him from a meeting, and loosed a barrage in Spanish that nearly flattened him, demanding that he go to speak with the judge immediately, file suit if it were necessary, because if they put Sabrina in a hospice for babies she would die no matter what. Willie got into gear and I went home to tremble and await results.

That night, very late, my husband returned bearing ten more years on his shoulders. I had never seen him so defeated, not even when he had to rescue Jennifer from a motel where she lay dying, cover her with his jacket, and take her to that hospital where she was received by the Filipino doctor. He told me that he had spoken with the judge, with the social worker, with the doctors, even with a psychiatrist, and that every one of them agreed that the baby's health was too fragile. "We can't take her on, Isabel. We don't have the energy to care for her or the strength to bear it if she dies. I'm not able to do this," he concluded, with his head in his hands.

A Gypsy at Heart

We had one of those fights that make history in a couple's lives and that deserve to be named—like the "Arauco war," which was what we in the family call a contest that kept my parents in battle mode for four months—but now, now that many years have gone by and I can look back on it, I concede that Willie was right. If there are enough pages, I will tell of other epic tourneys in which we have confronted each other, but I think that none was as violent as the battle over Sabrina; that one was a collision of personalities and cultures. I didn't want to hear his arguments; I was locked in a mute rage against the legal system, the judge, the social worker, Americans in general, and Willie in particular. We both stayed away from home as much as we could: Willie worked at his office far into the night, and

I packed a suitcase and went to stay with Tabra, who took me in without a fuss.

Tabra and I had known each other for several years; she was the first friend I made when I arrived in California. One day when she went to the beauty salon to have her hair tinted the beet red she was using then, the stylist commented that a week earlier a new client had come in and asked for the same color. We were the only two in her long professional career. She had added that the woman was a Chilean who wrote books, and mentioned my name. Tabra had read *The House of the Spirits*, and she asked the stylist to let her know the next time I was coming to the salon; she wanted to meet me. That happened fairly quickly because I had tired of the color sooner than I'd expected; I looked like a drowned clown. Tabra appeared with my book to be signed and was surprised to see that I was wearing earrings she had made. We were destined to hit it off, as the stylist said.

This woman who dressed in full Gypsy skirts, arms covered from wrist to elbow with silver bracelets, hair an impossible color, served as my model for the character Tamar in *The Infinite Plan*. I based Tamar on Carmen, a childhood friend of Willie's, and on Tabra, from whom I stole a personality and partial biography. Since Carmen inherited an impeccable moral rectitude from her father, she uses every opportunity to clarify that she never slept with Willie, a disclaimer that's entirely unnecessary except for people who have read my novel. Tabra's home—one story, wood, with high ceilings and large windows—was a museum for extraordinary objects from many corners of the planet, each with its own history: gourds used as penis shields from New Guinea, hairy masks from Indonesia, ferocious sculptures from Africa, dream paintings from the Australian aborigines. The property, which she shared with deer, raccoons, foxes, and the entire array of California birds, consisted of sixty acres of wild beauty. Silence, moisture, woodsy smells, a paradise obtained by dint of hard work and talent.

Tabra grew up in the bosom of southern fundamentalism. The Church of Christ was the one truth. Methodists did whatever they pleased, Baptists were damned because they had a piano in the church, Catholics didn't count—only Mexicans were Catholic and it wasn't proved that they had a soul—and the other denominations weren't worth talking about because their rites were satanic, as everyone knew. Alcohol, dancing, music, and swimming with anyone of the opposite sex were forbidden, and I think that was also true of tobacco and coffee, but I'm not sure. Tabra completed her education at Abilene Christian College, where her father taught, a sweet and open-minded professor enamored of Jewish and African-American literature, who navigated as well as he could through the censorship of the college authorities. He knew how rebellious Tabra was, but he had not expected her to elope with a secret boyfriend when she was seventeen, a Samoan student, the only person with dark skin and black eyes in that institution of whites. In those days, the youth from Samoa was still slender and handsome, at least in Tabra's eyes, and there was no doubt about his intelligence; up to that time he was the only Islander to have received a scholarship.

The couple ran away one night to another city, where the justice of the peace refused to marry them because interracial marriages were illegal, but Tabra convinced him that Polynesians are not Negroes, and furthermore, she was pregnant. Grumbling, the judge agreed. He had never heard of Samoa, and the hapless little mixed-blood creature she was carrying in her womb seemed reason enough to legitimize that disgraceful union. "I feel sorry for your parents, girl," he said instead of giving them his blessing. That same night the brand-new husband pulled off his belt and lashed Tabra until he drew blood because she had gone to bed with a man before she was married. The indisputable fact that he was that man did not in any way minimize her status as a whore. That was the first of countless beatings and rapes, which according to the church she had to endure because God did not approve of divorce, and that was her punishment for having married someone who was not of the same race, a perversion proscribed by the Bible.

They had a handsome son named Tangi, which in Samoan means "cry," and the husband took his small and terrified family back to his natal village. That tropical isle, where Americans maintained a military base and a detachment of missionaries, welcomed Tabra. She was the only white person in her husband's clan, and that afforded her a certain privilege, but it did not impede her husband's daily beatings. Tabra's new family consisted of some twenty dark-skinned giants, who lamented in chorus her pale, underfed appearance. Most of them, especially her father-in-law, treated her with affection and reserved for her the best bits of the communal dinner: fish heads with staring eyes, fried eggs enhanced by embryos, and a delicious pudding they prepared by chewing a fruit and spitting the pap into a wooden vessel they then set in the sun to ferment. Sometimes the women succeeded in picking up little Tangi and running to hide him from the fury of his father, but they were unable to defend his mother.

Tabra never grew accustomed to her fear. There were no rules regarding punishment; nothing she did or didn't do prevented the lashings. Finally, after one Homeric beating, her husband was sent to jail for a few days, a moment the missionaries seized to help Tabra and her son escape to Texas. The elders of her local church repudiated her. She couldn't find a decent job and the only person who helped her was her father. A divorce concluded that relationship and she didn't see her torturer again for fifteen years. By then, after many years of therapy, she was no longer afraid of him. Her former husband returned to the United States and became an evangelical preacher, a true scourge to sinners and unbelievers, but he never again dared bother Tabra.

In the decade of the '60s, Tabra could not bear the shame of the Vietnam War; she chose to leave the United States and travel with her son to different countries, where she made a living teaching English. In Barcelona, she studied jewelry-making and in the evenings strolled along the Ramblas to observe the Roma, from whom she drew inspiration for her Gypsy style. In Mexico, she was employed as an apprentice in a silversmith's workshop, and in a very short time she was designing and making her own jewelry. That, and only that, would be her calling for the rest of her life. With the defeat of the Americans in Vietnam, she returned to her country, and the era of the hippies found her, along with other penniless artists, in the colorful streets of Berkeley selling silver earrings, necklaces, and bracelets. During that period she slept in her beat-up car and used the university bathrooms, but her talent made her stand out among the other artisans and soon she could leave the street behind, rent a workshop, and hire her first helpers. A few years later, when I met her, she had a model enterprise located in a true Ali Baba cave replete with precious stones and objets d'art. More than a hundred persons were working with her, nearly all Asian refugees, some of whom had suffered the unimaginable, as was evident in their horrible scars and downcast eyes. They seemed to be very sweet people, although beneath the surface they must have hidden a volcanic desperation. Two of them, on two separate occasions, crazed by jealousy, bought a machine gun—taking advantage of the shops in this country where one can buy a personal arsenal—and killed the entire families of their wives. Then they blew out their own brains. Tabra had to attend those massive funerals and later had to "clean" the workplace with the necessary ceremonies so that bloody ghosts would not haunt the imagination of those left alive.

The face of Che Guevara, with his irresistible cha-
risma and his black beret pulled low on his forehead,
smiled from posters lining the workshop walls. During
a trip my friend made to Cuba with Tangi, she went
with the ex-chief of the Black Panthers to visit Che's
monument in Santa Clara. She brought with her the
ashes of a friend whom she had loved for twenty years,
without confiding it to anyone, and when they reached
the top of the memorial, she scattered them on the
wind. In that way she fulfilled his dream of traveling to
that mythic country. My friend's ideology is consider-
ably to the left of Fidel Castro.

"You're stuck in the mind-set of the '70s," I told
her once.

"And honored to be there," was her reply.

My beautiful friend's love affairs are as original as
her pythoness's clothing, her fiery hair, and her political
position. Years of therapy taught Tabra to avoid men
who might turn violent, as her Samoan husband had.
She swore that she would never let anyone beat her
again; nevertheless, it excites her to teeter on the edge
of the abyss. Only machos who look vaguely dangerous
or threatening attract her, and she doesn't like men of
her own race. Tangi, who had turned into a tall and
very handsome young man, did not want to hear a word
about his mother's sentimental difficulties. Some years

Tabra had as many as a hundred and fifty blind dates arranged through the personal ads in newspapers, but very few went further than the first cup of coffee. Following that, she chose more modern means, and now she is enrolled with several Internet agencies specializing in different types: "Single Democrats," with whom she at least has in common a hatred of Bush; "Amigos," which lists only Latinos, whom Tabra favors, but has the drawback that most of those men need a visa and try to convert her to Catholicism; and "Single Greens," who love Mother Earth but think that money isn't important and so don't work. She receives applications from very young studs with aspirations to be kept by a mature woman. Their photos speak worlds: dark, oily skin, naked torso, and the first inch or so of their fly unzipped to reveal the beginnings of pubic hair. The tone of the e-mail dialogues goes more or less like this.

> TABRA: Ordinarily I don't go out with men younger than my grandson.
> BOY: I'm more than old enough to fuck.
> TABRA: Would you talk like that to your own grandmother?

If someone of a more appropriate age for her shows up, he will turn out to be like the Democrat who lives with his mother and keeps his savings in silver ingots under the mattress. I'm not exaggerating: silver ingots, like the pirates of the Caribbean. It is strange that the Democrat in question would divulge on the first—and only—date information as private as where he hides his capital.

"Aren't you afraid to go out with strangers, Tabra? You might draw a criminal or a pervert," I commented when she had introduced me to a frightening type whose only allure was that he wore the beret of a Cuban *comandante.*

"It does make me think that I need a few more years of therapy," my friend admitted on that occasion.

Once she hired a painter to freshen up her walls, a fellow with a mane of black hair, something she really likes. On the basis of the hair Tabra invited him to lounge with her in the Jacuzzi. Bad idea; the painter began treating her like a husband. She would ask him to paint the door and he would answer, "Yes, dear," with visible irritation. One day he ran out of turpentine and announced that he needed an hour of meditation and a joint to get in contact with his inner space. By then Tabra was fed up with the black hair and told him that he had one hour to paint the interior space of the house and get the hell out of her life. He was no longer there when I arrived with my suitcase.

The first night, Tabra and I dined on fish soup, the only recipe my friend knows how to cook except for oatmeal with milk and sliced bananas. We got into the Jacuzzi, a slippery wood tub hidden among the trees; it had a sickening stench because an unfortunate skunk had fallen into it and simmered on a low flame for a week before it was discovered. There I unloaded my frustration like a bag of rocks.

"You want my opinion?" Tabra asked. "Sabrina won't ease your pain; grief takes time. You're very depressed, and you have nothing to offer that little girl."

"I can offer more than what she'll have in a foster home for very sick children."

"Then you'll have to do it alone, because Willie won't help you in this. I don't know how you plan to look after your son and your grandchildren, keep writing, and on top of that raise a little girl who needs two mothers."

Powerful Circle of Witches

A radiant Saturday dawned. Spring in Tabra's forest was already summer, but I didn't want to meet her and go for a walk, as we usually did on weekends. Instead, I called the five women who with me compose the circle of the Sisters of Perpetual Disorder. When I joined the group, they had already been meeting for several years to share their lives, meditate, and pray for people who were sick or in need of help. Now that I am one of them, we exchange makeup, drink champagne, stuff ourselves with bonbons, and sometimes go to the opera: spiritual practice alone is a little depressing to me. I had met them a year before, the day the physicians in California confirmed your diagnosis of no hope, Paula, exactly what I had been told in Spain. There was nothing that could be done, they said, you would never recover. I drove around keening in the car and I don't know how I ended up at Book Passage, my favorite bookstore, where I do a lot of my press interviews; they even keep a mailbox for me. There a Japanese lady almost as short as I am came over to me with an affectionate smile and invited me to have a cup of tea. She was Jean Shinoda Bolen, a psychiatrist and author of several books. I recognized her immediately because I had read her book on the goddesses that inhabit every woman, and how those archetypes influence personality. That was how I discovered that in me there was a jumble of contradictory deities that might be best not to explore. Though I had just met her, I told her what was happening with you. "We are going to pray for your daughter and for you," she told me. A month later she invited me to her "prayer circle," and that is how these new friends came to accompany me during your agony and death . . . and continue to comfort me today. For me it is a sisterhood sealed in heaven. Every woman in this world should have such a circle of friends. Each of us is witness to the others' lives; we keep secrets, help in difficulties, share experiences, and stay in almost daily contact by e-mail. However far I may be traveling, I always have my line to terra firma: my sisters of disorder. They are joyful, wise, and curious women. Sometimes curiosity can make one reckless, as in the instance of Jean herself, who in one spiritual ceremony felt an uncontrollable impulse, took off her shoes, and walked over red hot coals. Twice she passed through the fire, and emerged unhurt. She said it was like walking over little balls of Styrofoam; she felt the coals crunch and the rough texture of the burned wood beneath her feet.

During the long night at Tabra's, with the whispering of the trees and hooting of an owl, it occurred to me that the Sisters of Disorder might be able to help me. We met for breakfast in a restaurant filled with weekend sports enthusiasts, some in running shoes, others disguised as Martians to go cycling. We sat at a round table, always respecting the concept of the circle. We were six fiftyish witches: two Christians, an authentic Buddhist, two Jews by birth but semi-Buddhists by choice, and me, still undecided, all united by the same philosophy, which can be summed up in one sentence: Never do harm, and whenever possible do good. Between sips of coffee, I told them what was happening in my family, and ended with Tabra's words, which kept echoing in my head: Sabrina needs two mothers. "Two mothers?" repeated Pauline, one of the semi-Buddhists and a lawyer by profession. "I *know* two mothers!" She was referring to Fu and Grace, two women who had been together for eight years. Pauline went to the phone and made a call—at that time there were no cell phones. At the other end of the line, Grace listened to her description of Sabrina. "I'll talk to Fu and call you back in ten minutes," she said. Ten minutes . . . either they are unbalanced or they have hearts as big as the ocean to be able to decide something like that in ten minutes, I thought, but before the ten minutes were up, the restaurant's phone rang and Fu announced that she wanted to meet the baby.

I went to pick them up, driving along the rims of the hilltops in the direction of the ocean, a long, curving road that led to a poetic rural setting. Nearly invisible among pines and eucalyptus rose several Japanese-style wood constructions: the Buddhist Zen Center. Fu was tall and she had an unforgettable face: strong features, with a cocked eyebrow that gave her a questioning expression; she was dressed in loose, dark clothing, and her head was shaved like a draftee's. A Buddhist nun, she was the director of the center. She lived in a little dollhouse with her partner Grace, a physician who was irresistibly congenial and had the face of a mischievous child. In the car on the way back, I filled them in on the calvary that had been Jennifer's existence, the harm to the baby, and the specialists' dark prognosis. They did not seem daunted. We picked up Jennifer's mother, Willie's first wife, who knew Fu and

Grace because she'd attended ceremonies at the center, and the four of us drove on to the hospital.

In the section for newborns we were met by Odilia, she of the thousand curls, with Sabrina in her arms. She had already hinted, during an earlier visit, that she wanted to adopt Sabrina. Grace held out her arms and Odilia handed her the baby, who seemed to have lost weight and was shivering even more than before. But she was alert. Her large Egyptian eyes gazed into Grace's and then focused on Fu. I don't know what she told them in that first glance, but it was definitive. Without discussion, with a single voice, the two women declared that Sabrina was the little girl they had been waiting for all their lives.

I have been one of the Sisters of Disorder for several years now, and during that time I have witnessed a number of the marvels they have wrought, but none had such far-reaching effect as Sabrina. Not only did they find two mothers, they sorted out the bureaucratic tangle and facilitated Fu and Grace's being able to keep the child. By that time the judge had put his signature on the pertinent documents and Rebecca, the social worker, had declared the case closed. When we went to tell her that we'd found another solution, she informed us that Fu and Grace had no license, that they would have to take classes and go through special training to qualify as foster mothers; she added that they were not a traditional couple, and that they lived in another county and "the case" could not be transferred. Although Jennifer had lost custody of her daughter, her opinion still mattered, she added. "I'm sorry, but I don't have time to spend on something that's already been decided," she said. The list of obstacles continued, but I don't remember the details, only that at the conclusion of the interview, when we were about to leave in defeat, Pauline took Rebecca firmly by one arm.

"You have a very heavy caseload, and you are paid very little. You feel that your work is pointless, because in all the years you've been in this position you've not been able to save the wretched children who pass through this office," she said, looking deep into the woman's soul. "But believe me, Rebecca, you can help Sabrina. This may be your one chance to work a miracle."

The very next day Rebecca turned the bureaucracy upside down. She recovered all the paperwork and modified what was necessary, and she convinced the judge to sign again, to transfer the pertinent documents to another county, and to certify Fu and Grace as foster mothers . . . all in the blink of an eye. The same woman who the day before had been so indignant about our persistence had been converted into a radiant whirlwind who swept aside every obstacle and with the stroke of her magic pen determined Sabrina's fate.

"I told you, this child has an ancient and powerful soul," Odilia commented a couple of weeks later when she handed Sabrina over to her new mothers. "She touches people and they change. She has incredible mental power, and she knows what she wants."

So in the least expected manner, the monumental battle between Willie and me was resolved. We forgave each other, as much my dramatic accusations as his stubborn silence; we were able to put our arms around each other and weep with joy because that granddaughter had found her nest. Fu and Grace carried away their little mouse with the big wise eyes, and the circle of my friends set in motion the apparatus of their positive intentions to help her live. A photo of Sabrina sat on top of each home altar, and not a day went by that someone did not send up a thought for her. One of our sisters moved to another city, and we invited Grace to replace her in the group—after a period in which we verified that she had a sufficient sense of humor. In the Center of Zen Buddhism at least fifty persons prayed for Sabrina during their meditations and took turns rocking her while the two mothers struggled with her health problems that seemed never to end. During those first months it took five hours to give Sabrina two ounces of milk from an eyedropper. Fu learned to divine the symptoms of each crisis before it surfaced, and Grace, being a physician, had better resources than any of us.

"Are those women gay?" Celia, my daughter-in-law, asked. She had warned me more than once that she could not be beneath the same roof with someone whose sexual preferences did not coincide with hers.

"Of course."

"But one of them is a nun!"

"A Buddhist nun. She didn't take a vow of celibacy."

Celia said nothing more, but she was so impressed with Fu and Grace, whom she came to know very well, that she ended by questioning her own views. She had renounced religion long ago, and had no fear of the devil's cauldrons, but homosexuality was her strongest taboo. With time, however, she called them and asked forgiveness for the snubs of the past, and often visited them at the Zen Center, taking her children and her guitar to teach her new friends the rudimentary skills of motherhood and to cheer them with Venezuelan songs. Fervent environmentalists, the new mothers had planned to use cotton diapers for Sabrina, but before a month was out they had accepted the disposable ones

Celia brought as a gift. They would have had to be demented to go back to the old system of diaper pails and washing by hand. There is no washing machine in the center, everything is organic and difficult. The three became fast friends, and Celia began to show an interest in Buddhism, something that alarmed me because she tended to swing from one extreme to the other.

"It's a cool religion, Isabel. The only strange thing about those Buddha people is that they eat nothing but vegetables, like burros."

"I don't want to see you with your head shaved, or meditating in the lotus position, until you finish raising the children," I warned her.

Days of Light and Mourning

In September Celia gave birth to Nicole as calmly as she had welcomed Andrea sixteen months before. She endured ten hours of labor without a whimper, held by Nico, while I watched, thinking how my son wasn't any longer the boy I kept treating as if he were mine, but a man who with great composure had assumed the responsibility of a wife and three children. Celia, silent and pale, walked around between contractions, suffering before our helpless gazes. When she felt it was time, she lay on the bed, covered with sweat, trembling, and said something I will never forget: "I wouldn't trade this moment for anything in the world." Nico held her as the baby appeared, her head covered with dark fuzz, followed by shoulders and the rest of her body, wet, slippery, and streaked with blood, and once again

I experienced the epiphany I had the day Andrea was born and the unforgettable night you left us forever. Birth and death, Paula, are so similar . . . sacred and mysterious moments. The midwife handed me the scissors to cut the thick umbilical cord and Nico placed the baby on her mother's breast. Nicole was a plump packet of reinforced concrete that avidly latched onto the nipple as Celia talked to her in that unique tongue a mother, hazy from her ordeal and her sudden love, uses with her newborn. We had all been waiting for that child; she was a gift, and with her she brought a breath of redemption and joy. Pure light.

Nicole started screaming the instant she realized that she wasn't in her mother's womb any longer, and she never stopped for six months. Her shrieks peeled the paint off the walls and frayed the neighbor's nerves. Abuela Hilda, that beloved adopted grandmother who had been at my side for more than thirty years, along with Ligia, a Nicaraguan woman who had looked after you and whom I had hired to help with my grandchildren, rocked Nicole night and day, the only thing that quieted her for a few minutes. Ligia had left five children in her country and had come to work in the United States and support them from afar. It had been several years since she'd seen them and she had no hope of rejoining them anytime soon. For months and months those good women installed themselves and the baby in a rocker in my office, as Celia and I worked. I was afraid that from all that cradling and rocking my granddaughter's brain might be loosened from her skull and leave her impaired. Nicole calmed down the minute they began to give her powdered milk and soup. I think the cause for her despair was pure hunger.

In the meantime, Andrea was compulsively arranging her toys and talking to herself. When she got bored, she picked up her revolting *tuto*, announced that she was leaving for Venezuela, crawled into a cabinet, and closed the door after her. We had to bore a hole in that piece of furniture to provide a ray of light and breath of air, since my granddaughter could spend half a day without a word, locked in a space the size of a chicken coop. After Andrea's operation for strabismus, she had to wear glasses and a black patch that was changed every week from one eye to the other. So she wouldn't pull the glasses off, Nico dreamed up a contrivance made of six elastic bands and that many safety pins that crisscrossed over the top of her head. Some of the time Andrea tolerated it, but other times she flew into fits of exasperation and yanked the elastic bands until she managed to pull the whole thing down to diaper level. Incidentally, for a short while we had three children in diapers, and that is a *lot* of diapers. We bought them wholesale, and the most convenient system of changing the three was all at one time, whether they needed it or not. Celia or Nico would line up the opened-out diapers on the floor, lay the children on them, and wipe bottoms in a row, like an assembly line. They were able to do it with one hand while they talked on the telephone with the other, but I lacked their skill and always ended up plastered to my ears. The children were fed and bathed using the same one-two-three method. Nico got into the shower with them, soaped them, washed their hair, rinsed them off, and handed them out one by one for Celia to towel dry.

"You are a very good mother, Nico," I told him one day with sincere admiration.

"No, Mamá, I'm a good father," he replied, but I had never seen a father like him, and to this day I can't explain how he learned those skills.

At that time I was putting the final touches on my book *Paula*, struggling over the last pages, which were very painful for me. The memoir ended with your death, there was no other way it could end, but I was hazy on the details of that long night, it was swathed in a dense fog. I thought the room had been filled with people, and that I remembered seeing Ernesto in his Aikido whites, my parents, Granny, the grandmother who loved you so much, dead in Chile many years before, and others who could not possibly have been there.

"You were very tired, Mamá, and very sad. You can't remember the details," Nico excused me. "I don't remember them myself."

"What do those details matter? Write with your heart," Willie added. "You saw what we couldn't see. Maybe it's true that the room was filled with spirits."

I often opened the clay urn in which they had handed us your ashes. It is always on my writing table, the same table where my grandmother conducted her spiritist sessions. Sometimes I took out some letters and photographs of you before your illness, but I left those from your last year, when you were tied to your wheel-chair, inert. I have never touched those again, Paula. Still today, so many years later, I can't look at you in that state. I read the letters, especially that spiritual will with instructions to be followed in case of your death, which you had written on your honeymoon. You were only twenty-seven years old at the time. Why were you already thinking of death? I wrote that memoir with many, many tears.

"What's the matter?" Andrea asked sympathetically in her quasi-language, scrutinizing me with her cyclopean eye.

"Nothing, I just miss Paula."

"And why is Nicole crying?" she persisted.

"Oh, because she doesn't know any better," was the only answer I could think of.

Just as Alejandro had done before her, Andrea got it into her head that Paula was the only valid reason for crying. Since she could use only one eye at a time, she had no depth perception, everything was flat, and she gave herself some fearful bumps and bangs. She would get up off the floor streaming blood from her nose, with her eyeglasses all askew, and explain between sobs that she "missed" her Aunt Paula.

When I finished writing *Paula*, I realized that I had traveled a tortuous road and reached the end cleansed and naked. Those pages contained your luminous life and the trajectory of our family. The terrible confusion of that year of torment had dissipated. It had become clear that my loss was not exceptional but that of millions of mothers: the most ancient and most common suffering of humankind. I sent off the manuscript to people I'd mentioned in it; I felt that they should have the right to revise what I'd written about them. There weren't many because I'd left out several people who were close to you but that weren't essential to the main story. After they'd read it, everyone wrote me immediately, moved and enthusiastic, except for my best friend in Venezuela, Ildemaro, who adored you and thought that you would not enjoy seeing yourself exposed in that way. I'd had that same doubt, because it is one thing to write as catharsis, to honor a daughter you have lost, but quite a different thing to share your grief publicly. "You may be called an exhibitionist, or accused of using your tragedy to make money, you know how unkind people can be," my mother warned me; she was worried although she was convinced that the book should be published. To avoid any suspicion of the kind, I decided I would not touch a penny of the income from the book, if there was any; I would find an altruistic use for it, something that you would have approved of.

Ernesto was living in New Jersey, where he was working in the same multinational company that had employed him in Spain. When we brought you to our home, he asked for a transfer in order to be near you, but there was no position available in California and he had to accept the one he was offered in New Jersey—at least it was closer than Madrid. When he received the manuscript of the book, he called me, crying. It had been a full year since he'd been widowed but he still couldn't mention your name without breaking up. He encouraged me, using the charitable argument that you would like for the memoir to be published since it could console others who had losses and sorrow, but he added that he nearly hadn't recognized you in those pages. I had narrated the story from my narrow point of view. As your mother, there were areas of your personality and life that I knew nothing about. Where was Paula the impulsive lover, the finicky and bossy wife, the unconditional friend, the caustic critic? "I'm going to do something that Paula would kill me for if she knew," he told me, and three days later the mail brought me a large box containing the passionate love letters you two had exchanged for more than a year before you were married. It was an extraordinary gift that allowed me to know you better. With Ernesto's permission, I was able to include in the book actual lines you had written.

While I was polishing the final version of the book, Celia took complete charge of the office, wearing her blouse half buttoned so she could nurse Nicole at any minute. I don't know how she did it; she had three children, she was worn out, and she was carrying a profound sorrow. Her grandmother had died in Venezuela and she hadn't been able to tell her good-bye because her visa would not allow her to leave and then reenter the country. That grandmother, who treated everyone brusquely—except for Celia—had looked after her for three years when her parents were in the United States to work on doctorates in geology. When they returned,

Celia didn't know who those people were that she suddenly had to call Mamá and Papá. The pole star of her childhood had been her grandmother; she'd always slept with her, she told her her secrets, and only with her did she feel safe. A brother and sister were born after their return. Celia continued to be very close to her grandmother, who lived in an addition her parents built to their house. Celia's childhood in a strictly Catholic family could not have been easy given her rebellious and defiant character, but she submitted, and as an adolescent she had lived in an Opus Dei residence where the penances included self-flagellation and hair shirts with metal barbs. Celia says she didn't go to those extremes but she had to accept other rules meant to subdue the flesh: blind obedience, avoiding any contact with the opposite sex, fasting, sleeping on a board, spending hours on her knees, and other mortifications that were more frequent and more severe for women, since it is they who since the time of Eve have embodied sin and temptation.

Among the thousands of available young men in the university, Celia fell in love with Nico, who was precisely the opposite of what her parents had wanted in a son-in-law: he was Chilean, an immigrant, and an agnostic. Nico had been educated in a Jesuit school, but the day after he took his first communion he announced he did not plan to set foot in a church again. I met with the principal to explain that I would have to withdraw my son from the school, but the priest burst out laughing. "That won't be necessary, señora, we don't force anyone to go to mass. Your little guy is only nine years old, after all, and he may change his mind, don't you agree?" I had to admit that I didn't think so, because I know my son very well: he isn't one to make hasty resolutions. Nico completed his education at San Ignacio and fulfilled his word never to enter a church—with a few exceptions such as his religious wedding to Celia, and a few cathedrals he has visited as a tourist.

Celia had not been able to be with her grandmother as she died or weep at her death, for the truth is, Paula, you left no room for other mourning. Nico and I hadn't realized the magnitude of Celia's grief, partly because we didn't know the whole story of that period of her childhood and partly because Celia, taking pride in her fortitude, hid her pain. She buried that memory to cry over later, in the meantime performing the thousand tasks of maternity and matrimony, her job, learning English, and surviving in the new land she had chosen. During the few years we had shared, I had learned to love Celia, despite our differences, and after you were gone I clung to her as if she were another daughter. I was worried about the way she looked; her color was bad and she had no appetite, and she was suffering attacks of nausea as bad as the ones in the worst months of her pregnancy. Cheri Forrester, the family physician who had attended you, though you can't know that, said that Celia was physically worn down from having the three children in such close succession but that there was no physical cause for her vomiting, and that surely it was emotional; perhaps she was afraid that the porphyria would be repeated in one of their children. "If she doesn't improve, I'll have to keep her for a while in a clinic," she warned us. Celia kept quietly vomiting behind our backs.

A Peculiar Daughter-in-Law

Let me go back five years and remind you how your sister-in-law appeared in our lives. In 1988 I was living with Willie in California, you were studying in Virginia, and Nico, alone in Caracas, was finishing his last year at the university. He had announced during a telephone call that he was in love with one of his classmates and wanted to bring her to meet us, his feelings for her were serious. I asked straight out whether he wanted me to ready one room or two, and he answered, rather ironically, that from the point of view of the Opus Dei sleeping with a boyfriend would be unpardonable. Celia's parents were outraged by the sin of their traveling together without being married, even though she was twenty-five years old, and worse, that she was going to the home of a divorced Chilean atheist, Communist, and author of books banned by the church: me. That's all we need, I thought. Two rooms, for the present. Two of Willie's sons were living with us and my mother decided to come from Chile at just that time, so I improvised an army recruit's sleeping bag for Nico in the kitchen. My mother and I went to the airport to pick them up. We saw your brother, looking like the same clumsy adolescent, in the company of a person striding along with strong steps and carrying a bundle on her back that from a distance looked like a weapon, but turned out to be a guitar case. I suppose it was to annoy her mother, who had been a queen in some Caribbean beauty contest, that Celia walked like John Wayne, dressed in shapeless olive drab pants, mountain climbing boots, and a baseball cap pulled down over one eye. You had to look twice to discover how pretty she was; she had fine features, expressive eyes, elegant hands, broad hips, and an intensity in her gaze that was difficult to look away from. The young woman my son had fallen for came toward us defiantly, as if saying, "If you like me, fine, and if not, well, fuck you." She seemed so different from Nico that I was sure she was pregnant and they were planning a hasty wedding, but that turned out not to be true. It may have been that Celia just needed to get away from her surroundings for a while, and that feeling as if she were in a straitjacket, she had grabbed onto my son with the desperation of a person drowning.

When we got to the house, your brother announced that the sleeping bag in the kitchen would not be necessary because things had changed between them, so I put them in the same room. My mother took me by one arm and dragged me into the bathroom.

"If your son chose this girl, there's a good reason; your role is to love her and keep your mouth shut."

"But she smokes a pipe, Mamá!"

"It would be worse if she smoked opium."

It was easy for me to love Celia, even though I was a little shocked by her bold frankness and brusque ways—we Chileans tiptoe around a subject as if we were walking on eggs—and in less than half an hour she had expounded her ideas on inferior races, leftists, atheists, artists, and homosexuals, all of whom were depraved. She asked me please to let her know when anyone in any of those categories was coming to visit, she would prefer not to be present. That night, however, Celia kept us laughing with off-color jokes we hadn't heard since the easy-going days in Venezuela, where happily the concept of "politically correct" does not exist and you can make jokes on any subject you choose, and then took her guitar from its case and sang to us in an engaging voice the best songs in her repertoire. We were captivated.

———————

Shortly after, Celia and Nico were married in Caracas, in a long, drawn-out ceremony during which you threw up in the bathroom—I think out of jealousy because you were losing exclusive rights to your brother—and from which my family took early leave as it seemed that we didn't fit in. We knew almost no one there and Nico had warned us that his bride's relatives did not feel kindly toward us. We were political refugees and had escaped Pinochet's dictatorship, and therefore we were probably Communists with no social standing, or money, and we didn't belong to the Opus Dei. We weren't even practicing Catholics. The newlyweds moved into the house I had bought when I was still living in Caracas, though it was too big for them, and Alejandro, your first nephew, was born a year later. I shot out of San Francisco, flew hour after hour—counting the minutes and shivering with anticipation—and in Caracas took into my arms a newborn smelling of mother's milk and talcum powder while out of the corner of my eye I studied my daughter-in-law and my son with growing admiration. They were two little kids playing with dolls. Your brother, who only a short time before had been an irresponsible boy who risked his life climbing mountains and swimming with sharks in the open sea, now was changing diapers, warming bottles, and cooking pancakes for breakfast, side by side with his wife.

The one worry in the lives of this couple was that the criminal element in Caracas had targeted their house. They had stolen things countless times; they had taken three cars right out of the garage, and now alarms, bars on the windows, and electrically wired grilles that would roast a careless cat that brushed them with a whisker had no effect. Every time they came home, Celia stayed in the car, holding the baby and with the motor running, while Nico, pistol in hand, got out and, the way you see in films, checked the house from top to bottom to be sure that some cold-blooded intruder wasn't hiding somewhere. They lived in fear, which worked out well for me since it made it easier to convince them to move to California, where they would be safe and could count on our help. Willie and I fixed up a wonderful little bohemian garret with a tower overlooking the panorama of San Francisco Bay. It was on a third floor and there was no elevator, but they were young and strong and they could fly up and down the stairs with baby paraphernalia, shopping bags, and the garbage. I waited for them with all the nervous anticipation of a bride-to-be, prepared to squeeze the last drop out of my new status as a grandmother. More than once I wound the little music boxes and the mobiles hanging from the ceiling of Alejandro's room and sang in whispers the nursery songs I had learned when you and your brother were little. The wait seemed eternal, but time inevitably passes and finally they arrived.

At first my friendship with Celia stumbled along in fits and starts. Mother- and daughter-in-law came from widely divergent ideologies, but if we had any idea of bickering over differences, life eliminated what might have been bad blood with a few knocks to the head. Soon we forgot any germ of discord and concentrated on the demands of raising a child—and then two more—while adapting to a new language and our situation as immigrants. Although we didn't know it then, a year later we would have our most brutal test: caring for you, Paula. There would be no time for foolishness. Celia very quickly cut the strings that bound her to her fanatic Catholicism and began to question other precepts that had been hammered into her head in her youth. As soon as she realized that in the United States she was not considered white, the racism faded away, and her friendship with Tabra had swept away her prejudices against artists and persons with leftist leanings. Of homosexuals, however, best not to speak. She hadn't as yet met Sabrina's mothers.

Nico and Celia enrolled in an intensive English course, and the happy task of taking care of my grandson fell to me. As I wrote, Alejandro crawled around on the floor, kept captive by the dog gate we installed at the door. If he got tired, he would stick his pacifier in his mouth, drag over his pillow, and fall asleep at my feet. When it was time to eat, he tugged a few times on my skirt to pull me from the trance I tend to sink into when I write, and I would distractedly reach for his bottle and he would drink it without a sound. Once he unplugged the cable of my computer and I lost forty-eight pages of my new book, but instead of throttling him, as I would have any other mortal, I ate him up with kisses. The pages weren't good anyway.

My happiness was nearly complete; you were all that was lacking. In 1991 you had recently married Ernesto and were living in Spain, but you already had plans to move to California, where we would all be together. On December 6 of that same year, Paula, you went to the hospital with a stomachache and a cold too long ignored. You don't know what happened there. Hours later you were in intensive care in a coma, and five eternal months would go by before you were handed over to me in a vegetative state, with severe brain damage. You were breathing; that was your only sign of life. You were paralyzed and your eyes were black pools that no longer reflected light, and in the months that followed, you changed so much that it was difficult to recognize you. With the help of Ernesto, who refused to admit that in reality he was already a widower, we brought you to my home in California on a harrowing flight across the Atlantic and the continental United States. Then Ernesto had to leave you with me and go back to his job. I never imagined that the dream of having my daughter close to me would come true in such tragic fashion. Celia was near the time of giving birth to Andrea. I remember her reaction when they lowered you from the ambulance on a stretcher. She clung to Alejandro, retreated, trembling, her eyes wide with shock, as Nico paled and took a step forward; he leaned down to give you a kiss, as his tears rained down on you. For you this world ended on December 6, 1992, exactly one year after you entered the hospital in Madrid. Days later, when we scattered your ashes in a nearby forest, Celia and Nico informed me that they planned to have another child. Nicole was born ten months later.

Green Tea for Sadness

Willie realized with desperation that Jennifer was gradually committing suicide. An astrologer had told him that his daughter was "in the house of death." According to Fu, there are souls who unconsciously try to achieve divine ecstasy by way of the expeditious path of drugs; maybe Jennifer needed to escape the gross reality of this world. Willie believes that he has transmitted bad genes to his children. His great-great-grandfather had arrived in Australia with shackles on his legs, covered with pustules and lice, one among a hundred and sixty thousand wretches the English sent to that land to serve out their sentences. The youngest of the convicts, sentenced for stealing bread, was nine years old, and the eldest was an old lady of eighty-two who'd been accused of stealing two pounds of cheese and who hanged herself a few days after her ship docked. Willie's ancestor, accused of who knows what rubbish, had not been hanged because he was a knife-sharpener. In those years, having a trade or knowing how to read meant that instead of being hanged you were sent to Australia. The man was among the strong ones who survived, thanks to his ability to absorb suffering and alcohol, an aptitude he passed on to nearly all his descendants. Very little is known about Willie's grandfather, but his father died of cirrhosis. Willie himself spent decades of his life without tasting a drop of alcohol because it triggers his allergies, but if he started, the amount gradually crept up. I have never seen Willie drunk. Before he reaches that point, he chokes as if he had swallowed a fistful of hair and is rendered inoperative by a ferocious headache; we both know, however, that if it weren't for those blessed allergies, he would have ended up like his father. Only now, after reaching sixty, has he learned to limit himself to a single glass of white wine and feel satisfied. It is said that we cannot duck our heritage, and his three children—all drug addicts—seem to confirm that. They do not have the same mother, but in the family lines of his first and second wives there is also addiction, handed down from their grandfathers. The only child who has never waged war against Willie is Jason, his second wife's son by another man, whom he loves as if he were his own. "Jason doesn't have my blood; that's why he's normal," Willie tends to comment in the tone of someone reporting a natural event like the tides or the migration of wild ducks.

When I met him, Jason was a boy of eighteen, with a lot of talent for writing but lacking discipline, though I was sure that sooner or later he would acquire it. That's what it takes to deal with the rigors of life. He planned to be a writer some day, but in the meantime he was contemplating his navel. He would write two or three lines and come running to ask me if maybe there was potential there for a story, but it never went any further than that. I myself pushed him out of the house to go study at a college in southern California, where he graduated with honors, and when he returned to live with us he brought his girlfriend, Sally. Jason's biological father had a volatile temperament that tended to explode with unpredictable consequences. When Jason was only a few weeks old, there was an accident that was never clarified. His father said that the baby had fallen off the changing table, but his mother and the physicians suspected that he had been struck on the head, denting his skull. They had to operate, and by some miracle the baby came out sound—after spending a lot of time in the hospital while his parents were getting a divorce. From the hospital he was passed to the care of the state; then his mother took him to live with an aunt and uncle who according to Jason were true saints, and finally she brought him to California. When he was three, the boy went to live with his father because it seems that the building where his mother lived did not accept children. What kind of building would that be? When she married Willie, she reclaimed the boy. Later, when they were divorced, the child picked up his belongings and without hesitation went to live with Willie. In the meantime, his biological father made sporadic appearances, and on occasion again mistreated him—until Jason was old enough, and had the physical presence to defend himself. One night of heavy drinking and recriminations in his father's cabin in the mountains, where they'd gone for a few days' vacation, the man starting hitting Jason, who had promised himself he would never again allow himself to be victimized, and he responded with all the fear and rage that had accumulated for years, and used his father's face for a punching bag. Horrified, he drove several hours through a stormy night to get home; his shirt was stained with blood and he was nearly sick with guilt. Willie congratulated him; it was time to lay out the ground rules, he said. That distressing incident established an accord of respect between father and son. The violence was never repeated, and now they have a good relationship.

That year of mourning, of too much work, of financial difficulties and problems with my stepchildren, was undermining the foundation of my relationship with Willie. There was too much chaos in our lives. I wasn't adapting to the United States. I felt that my heart was growing cold, that it wasn't worth the effort to keep on rowing against the current; the energy needed to keep us afloat was disproportionate. I thought about leaving, running away, taking Nico and his family to Chile, where at last, after sixteen years of military dictatorship, democracy had been restored, and where my parents lived. Get a divorce, that's what I have to do, I would mutter under my breath, but I must have said it aloud more than once because Willie cocked an ear when he heard the word *divorce.* He had gone that route twice before and was determined there would not be a third time, and he pressed me to go with him to see a counselor. For years I had made fun of Tabra's therapist, a wild-haired alcoholic whose counsel consisted of exactly the same package of platitudes I would have given without charge. In my opinion, therapy was a mania of North Americans, a very spoiled people unable to tolerate the normal difficulties of life. When I was young, my grandfather had instilled in me the stoic notion that life is hard, and when facing a problem there is nothing to do but grit our teeth and keep going. Happiness is pure kitsch; we come into the world to suffer and learn. Fortunately, the hedonism of Venezuela shook my belief in my grandfather's medieval precepts and gave me permission to enjoy myself without feeling guilty. At the time of my youth in Chile, no one visited a therapist—except for certifiable lunatics and Argentine tourists—so I strongly resisted Willie's suggestion, but he was so persistent that finally I gave in and went with him. More accurately, he took my arm and dragged me there.

The psychologist turned out to look like a monk with a shaved skull, who drank green tea and sat through most of the session with his eyes closed. In Marin County, at any time of day, you see men riding bicycles, jogging in shorts, or savoring a cappuccino at little sidewalk tables. "Don't these people work?" I once asked Willie. "They're all therapists," he'd answered. Which may be why I felt so skeptical when we met with Bald Head, who was really very wise, as I soon found out. His office was a bare room painted a kind of pea green and decorated with a large wall hanging—a mandala, I think they're called. Willie and I sat cross-legged on cushions on the floor while the monk sipped his Japanese tea like a little bird. We began talking and soon a whole avalanche was unleashed. Willie and I each tried to get our stories in first, to tell him about what had happened with you, about the terrifying life Jennifer lived, about Sabrina's fragility and a thousand other problems, and my desire to say, The hell with it, and disappear. The tea-sipper listened without interrupting, and when only a few minutes were left to end the session, he opened his heavy lidded eyes and looked at us with an expression of genuine sorrow. "What sadness there is in your lives!" he murmured. Sadness? Actually, that hadn't occurred to either of us. All the air blew out of our rage in an instant, and deep in our bones we felt a grief as vast as the Pacific Ocean, a pain we hadn't wanted to admit out of pure and simple pride. Willie took my hand, pulled me to his cushion, and we hugged each other tight. For the first time, we admitted that our hearts were broken. It was the beginning of our reconciliation.

"I am going to suggest that you do not mention the word *divorce* for an entire week. Can you do that?" the therapist asked.

"Yes," we answered simultaneously.

"And could you do it for two weeks?"

"Three, if you want," I said.

That was our agreement. For three weeks we focused on solving everyday emergencies, and never spoke the forbidden word. We were living in a state of crisis, but the allotted time went by, then a month, then two, and the truth is that we never again spoke of divorce. We went back to the nightly dance that from the beginning had been so natural: sleeping so tightly embraced that if one turns the other adjusts, and if one rolls away, the other wakes. Between countless cups of green tea, the shaved-head psychologist led us by the hand over the rough terrain of those years. He counseled me to "stay in my trench" and not interfere in the problems with Willie's children, who in truth were the principal cause of our fights. So Willie gives a new car to his son, who has recently been expelled from school and is floating around in a cloud of LSD and marijuana? Not my problem. And he crashes it against a tree two weeks later? I stay in my trench. Willie buys him a second car, which he also destroys? I bite my tongue. Then his father rewards him with a van and explains to me that it is a safer, stronger vehicle. "Of course. That way when he runs over someone, at least he won't leave him wounded, he'll kill him outright," I reply with glacial calm. I lock myself in the bathroom, take an icy shower, and recite all the curse words in my Spanish repertoire, then spend a few hours making necklaces in Tabra's workshop.

The therapy was very helpful. Thanks to it and my writing, I survived an assortment of trials, though I did not always come out the winner, and my love for Willie was saved. Fortunately, though, the family melodrama continued, because if not, what the devil would I write about?

A Girl with Three Mothers

Jennifer was allowed to see Sabrina in supervised visits every two weeks, and with every one I could see how Willie's daughter's health was deteriorating. She looked worse every time I saw her, as I wrote my mother and my friend Pía. In Chile they both had made donations to Padre Hurtado's foundation; he is the only Chilean saint that even Communists venerate because he can work miracles, and they were praying for Jennifer to be cured of her addictions. In truth, only divine intervention could help her.

And here I want to pause briefly to introduce Pía, my forever friend, the woman who is like my Chilean sister, whose loyalty has never wavered, not even when we were separated by my exile. Pía comes from a very conservative Catholic family that celebrated the military coup of 1973 with champagne, but I know that on at least two occasions she hid victims of the dictatorship in her house. It is rare that we speak about politics, for we don't want anything to come between us. After I took my small family to Venezuela, we kept in touch by letter, and now we visit each other in Chile and in California, where she likes to come for vacations, and so we have kept alive a friendship that by now has a diamantine clarity. We love each other unconditionally and when we're together we create four-handed paintings and giggle like schoolgirls. Do you remember that Pía and I used to joke about how one day we would be two merry widows and would live together in a garret, gossiping and making our crafts? Well, Paula, we don't talk about that anymore because Gerardo, her husband, the kindest and most guileless man in this world, died one morning like any other when he was supervising work in one of his fields. He sighed, bowed his head, and went to the other world without a good-bye. Pía can't be consoled even though she is surrounded by her clan: four children, five grandchildren, and scores of relatives and friends with whom she is constantly in touch, as is the custom in Chile. She devotes herself to charities of every sort, takes care of her family, and works with her oils and brushes in her free time. In moments of sadness, when she can't stop crying over Gerardo, she closes her door and creates small works of art with scraps of cloth, including icons embroidered with beads and precious stones that look as if they'd come from the treasure troves of ancient Constantinople. This Pía who loved you so much had a tiny chapel built in her garden and planted a rose in your memory. There beside that luxuriant rosebush she talks with Gerardo and you, and often prays for Willie's children and for his granddaughter.

Rebecca, the social worker, organized the routine for Sabrina's visits with her mother. It wasn't easy, since the judge had ordered that Jennifer and her companion should not meet the foster mothers or learn where they lived. Fu and Grace would meet me in the parking lot of some mall and give me the child, with diapers, toys, bottles, and the rest of the paraphernalia babies need. I would drive her, in one of the seats I kept in my car for my grandchildren, to City Hall, where I would meet Rebecca and a policewoman—always a different one, though they all had an air of professional boredom. While the uniformed woman watched the door, Rebecca and I waited in a nearby room, enchanted with Sabrina, who had become very beautiful and very alert; she did not miss a single detail. She had caramel-colored skin, the fuzz of a newborn lamb on her head, and the amazing eyes of a houri. Sometimes Jennifer would show up for the meeting, sometimes not. When she did appear, with a bad case of jitters—a fox being chased by hounds—she never stayed more than five or ten minutes. She would pick up her daughter, but then feeling her light in her arms, or hearing her cry, she seemed confounded. "I need a cigarette . . . ," she'd say, and she would hurry out and often not come back. Rebecca and the police officer would take Sabrina and me back to my car, and I would drive to the parking lot where the two mothers were anxiously waiting. I think that for Jennifer those harried visits must have been a torment; she had lost her daughter, and not even her relief at knowing she was in good hands could console her.

These strategic appointments had been under way for about five months when Jennifer was again taken to the hospital, this time with an infection in her heart and another in her legs. She showed no signs of alarm, but simply told us that it had happened before. Nothing serious, she insisted, but the doctors were not so sanguine. Fu and Grace decided that they were tired of hiding and that Jenny had the right to know the women who were looking after her daughter. I went with them to the hospital, ignoring legal protocol. "If the social worker finds out, you all will be in a jam," warned Willie, who thinks like a lawyer and still did not know Rebecca well.

Jennifer was a pitiable sight; you could count her teeth through the translucent skin of her cheeks, her hair was a tangled doll's wig, and her hands were blue and the nails black. Her mother was also there, horrified to see her daughter in that state. I think she had accepted the fact that Jennifer would not live much longer, but was hoping at least to reconnect with her before the end. She thought that they would talk and make peace in the hospital—after so many years of hurting each other—but once more her daughter would run away before the medications could take effect. Our difficulties made Willie's first wife and me very close; she had suffered with her children—both of them were addicted to drugs—and I had lost you, Paula. She had been divorced from Willie for more than twenty years, and both of them had remarried. I don't think there was any lingering bad feeling, but if there was, the arrival of Sabrina in their lives had redeemed it. The attraction that had brought Willie and her together in their youth had turned to mutual disillusion shortly after their marriage, and had ended ten years later in divorce. Except for their children they had nothing in common.

During the years they were married, Willie was entirely dedicated to his career, determined to be successful and make money, and his wife felt abandoned and often fell into deep depressions. It was, furthermore, their fate to have lived in the turbulence of the '60s, when customs were greatly relaxed in this part of the world. Free love was in vogue, couples swapped partners as a form of entertainment, at parties people bathed naked in Jacuzzis, and everyone drank martinis and smoked marijuana, while the children ran wild through the middle of it all. Those experiments left in their wake a multitude of easily predictable destroyed marriages. But Willie says that that wasn't the cause of the break. "We were like oil and water; we didn't blend together, that marriage couldn't last." At the beginning of my relationship with Willie, I asked him whether our arrangement was going to be "open"—a euphemism for mutual infidelity—or monogamous. I needed to have that clarified because I have neither time nor inclination to spy on a fickle lover. "Monogamous," he replied without hesitation. "I've tried the other formula and it's a disaster." "That's good, but if I catch you in a little peccadillo I'll kill you, your children, and the dog. Do you hear what I'm saying?" "Perfectly." I myself have respected our deal with more decency than might be expected of a person of my character, and I suppose he has done the same, but I wouldn't stake my life on it.

Jennifer took her baby and held her to her squalid breast, as she thanked Fu and Grace over and over.

Both of those women have the gift of investing everything they touch with humor, calm, and beauty. They breached Jennifer's defenses—something no one had ever accomplished before—and prepared to accept her with all their compassion, which is considerable. Thus a sordid drama was transformed into a spiritual experience. Grace stroked Jennifer, smoothed her hair, kissed her forehead, and assured her that she could see Sabrina every day if she wished; she herself would bring her, and when Jennifer was released from the hospital she could visit the baby at the center. She told how intelligent and lively Sabrina was, how she was beginning to drink milk without difficulty, but did not mention any of her serious health problems.

"Don't you think Jennifer should know the truth, Grace?" I asked as we left.

"What truth?"

"That if Sabrina keeps growing weaker at this rate . . . her white cells—"

"She's not going to die. I can swear to that," she interrupted with calm conviction.

That was the last time we would ever see Jennifer.

On May 25, 1994, we celebrated Sabrina's first birthday at the Zen Center, in a circle of some fifty barefoot people, some wearing the loose robes of medieval pilgrims, some with shaved heads, and some with that

expression of suspicious placidity that earmarks vegetarians. Celia, Nico, their little ones, Jason, with his
girlfriend, Sally, and the rest of the family were there.
The only woman wearing makeup was me, and the
only man with a camera was Willie. In the center of
the room, amid a riot of balloons, several children were
playing around a monumental organic carrot cake.
Sabrina—crowned queen of Ethiopia by Alejandro—
dressed as a gnome, with metallic star stickers on her
forehead and with a yellow balloon tied to her belt so
she could be seen and not stepped on, was passed from
arm to arm and from kiss to kiss. Compared to my
granddaughter Nicole, who was as solid as a koala bear,
Sabrina looked like a soft little doll, but in that one year
she had defied nearly all the fatalistic prognoses of the
doctors; she was now able to sit up and she was trying
to crawl, and she could identify all the residents of the
Zen Center. The invited guests, one by one, introduced themselves. "I am Kate. I take care of Sabrina
on Tuesdays and Thursdays." "My name is Mark and I
am her physical therapist." "I am Michael, a Zen monk
for thirty years, and Sabrina is my teacher. . . ."

Little Everyday Miracles

December 6, 1993, was the first anniversary of your death. I wanted to remember you as beautiful, unassuming, content, dressed as a bride or holding a black umbrella and leaping over puddles in the rain in Toledo, Spain. But at night, in my bad dreams, I was assaulted by the most tragic images: your hospital bed, the hoarse sound of the respirator, your wheelchair, the handkerchief we used to cover the hole of the tracheotomy, your clenched hands. I had prayed so many times to die in your place, and later, when that exchange was impossible, prayed to die after you did, that in all fairness I should have been seriously ill. But dying is very difficult, as you know and as my grandfather told me shortly before he completed a century of living. A year had gone by since your death and I was still alive, thanks to my family's affection and the magical needles and Chinese herbs of my wise Japanese friend Miki Shima, who had been with you and me during those long months when you were saying goodbye. I don't know what effect his remedies had on you, but his tranquil presence and spiritual messages kept me going week after week. "Don't say you want to die, that makes me so sad that *I* want to die," my mother reproached me once when I hinted at that in a letter. She was not my only reason for living; I had Willie and Nico and Celia, and those three grandchildren who often woke me with their grubby little hands and slobbery kisses, smelling of sweat and pacifiers. At night, all in the same bed snuggled close together, we would watch frightening videos of dinosaurs devouring the actors. Alejandro, four years old, would take my hand and tell me not to be afraid, it was all a big fib; afterward the monsters vomited up the people whole because they didn't chew them.

On the morning of that anniversary I took Alejandro to the forest we now call "Paula's forest." That's rather presumptuous of us, daughter, it *is* a state park. It was raining, and very cold; our feet sank into the mud, the air smelled of pines, and a sad winter light was filtering through the treetops. My grandson ran ahead of me, toes out and arms flapping like a duck. As soon as we neared the stream—tumultuous in winter—where we had scattered your ashes, he immediately recognized it.

"Paula was sick yesterday," he said. For him anything in the past was yesterday.

"Yes. She died."

"Who killed her?"

"It wasn't like television, Alejandro. Sometimes people get sick and just die."

"Where do dead people go?"

"I don't know exactly."

"She went down there," he said, pointing to the stream.

"Her ashes went with the water, but her spirit lives in this forest. Isn't this a great place?"

"No," he decided. "It would be better if she was living with us."

We stayed a long time remembering you in that green cathedral, where we could feel you, tangible and present, like the cold breeze and the rain.

That evening the family, including Ernesto, who'd come from New Jersey, and a few friends got together at our home. We sat in the living room and celebrated the gifts you had given us during your lifetime and the gifts you continued to give us, such as the births of the grandchildren Sabrina and Nicole, and the incorporation into our tribe of the mothers Fu and Grace, along with Sally. A humble white candle with a hole in the center presided over the altar we had improvised to hold your photographs and mementos.

The year before, three days after your death, I had met with the Sisters of Disorder at one of their homes, as we always did on Tuesday; we made a circle around six new candles. I was numb with grief over your departure. "I have this burning in the center of my body, like a fire in my womb," I told them. We joined hands, closed our eyes, and my friends directed toward me their affection and their prayers, to help me endure the pain of those days. I asked for a sign, an indication that you had not disappeared into nothingness forever, that your spirit existed somewhere. Suddenly I heard Jean's voice. "Look at your candle, Isabel!" My candle was burning in the center. "There's your fire in the womb," Jean added. We waited. The flame melted the wax and formed a hollow in the middle of the candle, but it did not bend or split apart. Just as it had spontaneously caught fire, instants later the flame went out. The candle was hollowed out, but erect, and I took that as the sign I was waiting for, a wink you'd sent me from another dimension: the raw burn of your death would not break me. Nico later examined the candle and couldn't find the cause of that strange flame in the center; maybe the candle was defective, or had a double wick that lighted from a spark. "Why do you want an explanation, Mamá?" he asked. "What matters is that you received the sign you asked for, that's enough." I suppose he wanted me to be content, because given his healthy skepticism, I don't think he believed it was a miracle.

Fu explained to all of us that we lighted incense because the smoke rises like our thoughts, and the light of the candles represents wisdom, clarity, and life. Flowers symbolize beauty and continuity; they die but they leave seeds for other flowers, just as our seeds survive in our grandchildren. Each person shared some sentiment or memory. Celia, the last to speak, said, "Paula, remember that you have two nieces and a nephew, and you must take very good care of them, for they may have porphyria too. Remember to watch and see that Sabrina has a long and happy life. And remember, too, that Ernesto needs another wife; so go ahead and find him a girlfriend."

To end our evening we mixed earth with a pinch of the ashes I had saved from your cremation, and planted a little tree in a pot, with the idea that as soon as it set its roots we could put it out in our garden or in your forest.

Cheri Forrester, our compassionate doctor, was also there, along with Miki Shima, who days before had cast the I Ching sticks for me. What they had said was: *"The woman has patiently tolerated the desolate earth; she*

*is crossing the river, barefoot and with determination;
she counts on people who are at a distance, but does
not have companions; she must walk alone through the
middle crossing."* I thought it was very clear. Dr. Shima
said that he had received a message from you. "Paula
is fine, she is moving along her spiritual path but she
looks after us and is present among us. She says she
does not want us to keep weeping over her, she wants
to see us happy." Nico and Willie exchanged meaning-
ful glances; they do not fully believe in this fine man.
They argue that he can't prove anything he says, but
that night I had no doubt that it was your voice because it was so similar to the message you left in your
will. *"Please, don't be sad. I am still with all of you,
just closer than before. Later on we will be reunited
in spirit, but in the meantime we will be close as long
as you remember me. Don't forget that we, we spirits, most effectively help, accompany, and protect those
who are content."* That is what you wrote, daughter.
Cheri Forrester cried and cried because her mother
died at your age, and from what she said, you two were
very much alike physically.

I had intended on that memorable day to put the final
word to the manuscript of the book and offer it to you
as a gift. Fu blessed the bundle of papers tied with a red
ribbon, and we all toasted with champagne and shared

a chocolate cake. There was deep emotion, though it wasn't an evening of mourning but, rather, a quiet ceremony. We were celebrating that at last you were free after having spent so much time as a prisoner.

Sadness. As the therapist had pointed out, there was sadness in both Willie's life and mine. It was not a paralyzing emotion but an awareness of the losses and difficulties that colored our reality. We often had to adjust our burdens in order to go forward without falling. Everything was disorganized; we had the feeling we were living at the center of a storm, boarding up doors and windows so the winds of misfortune did not level everything.

Willie's office was operating on credit. He accepted hopeless cases, spent more than he earned, maintained a herd of useless employees, and was entangled in a number of tax wrangles. He was a terrible administrator, and Tong, his loyal Chinese accountant, could not control him. My presence in his life brought stability because I could help with expenses in emergencies, run the house, check the bank balances, and do away with most of the credit cards. He moved his San Francisco office to a Victorian house I bought in Sausalito, the most picturesque town on the bay. The property had been built around 1870 and boasted a notable pedigree:

it was the first brothel in Sausalito. Later it was converted into a church, then a chocolate cookie factory, and finally, a complete ruin, it passed into our hands. As Willie said, it kept sliding down the social ladder. It sat among sick, centuries-old trees that threatened to fall onto the neighbors' houses with the first gale. We were forced to cut down two of them.

The executioners arrived dressed like astronauts; they climbed up the trees with saws and axes, swung from the branches on ropes, and proceeded to draw and quarter their victims, who bled to death quietly, as trees do. I had to run away, unable to witness that massacre any longer. The next day we didn't recognize the house. It was naked and vulnerable, its wood devoured by time and termites, the shingles twisted, the shutters dangling. The trees had hidden the degree of deterioration; without them the house resembled a decrepit courtesan. Willie enthusiastically rubbed his hands. In some previous life he had been a builder, one of the ones who construct cathedrals. "We are going to make this house as beautiful as it was when it was young," he said, and set off in search of the original plans to return it to its Victorian grace. He succeeded magnificently and, despite the profanation of tools, its walls still hold the French perfume of the whores, the Christian incense, and the chocolate of cookies.

In the same rooms where the long-ago ladies of the night made their clients forget their sorrows, Willie today combats the uncertainties of the law. In what was formerly the carriage house, I clashed with my literary ghosts for years, until I had my own *cuchitril* at our house, where I now write. Using the move as an excuse, Willie got rid of half of his employees and then was able to choose his cases more carefully; his office, nonetheless, was still chaotic and not profitable. "However much you bring in, more goes out. Add it up, Willie. You're working for a dollar an hour," I pointed out. Willie was never fond of keeping tabs, but Tong, who had worked for him for thirty years and had more than once saved him from bankruptcy by a hair, agreed with me.

I grew up with a Basque grandfather who was very cautious with money, and then with my Tío Ramón, who barely survived on his meager salary. My stepfather's philosophy was, "We are filthy rich," no matter that out of necessity he had to be very prudent with expenditures. He proposed to live life in grand style, though he had to stretch every cent of his paltry pay as a public servant to maintain his four children and my mother's three. Tío Ramón would divide his month's salary and put the money to cover our basic needs, counted and recounted, into four envelopes; each had to last a week.

If he managed to save a little here and a little there, he would take us to get ice cream. My mother, who was always considered a very stylish woman, made her own dresses, transforming them again and again. They had an active social life, unavoidable for diplomats, and she had a basic gray silk evening gown to which she added and removed sleeves, belts, and bows, so that in photographs of the day she always appears in a different dress. It never passed through their minds to go into debt. Tío Ramón gave me my most useful guidelines for living, as I discovered in therapy as a mature woman: a selective memory for remembering the good things, a logical prudence to avoid doing anything to ruin the present, and defiant optimism for facing the future. He also instilled a spirit of serving, and taught me not to complain because that will ruin your health. He has been my best friend; there is nothing I haven't shared with him. Because of the way he and my mother brought me up, added to the alarms of exile, I have a peasant mentality when it comes to money. If it were up to me, I would hide my savings under the mattress, as Tabra's former suitor did with his bars of silver. The way my husband went through money horrified me, but every time I stuck my nose in his business, it caused a battle.

After the manuscript of *Paula* was sent to Spain and had safely arrived in the hands of Carmen Balcells, I was overcome by a profound weariness. I was extremely busy with family, travels, lectures, readings, and the bureaucracy of my office, which had been growing until it had reached terrifying proportions. Time refused to do my bidding; I was circling around in the same spot like a dog chewing its tail, and not producing anything worthwhile. I kept trying to write. I had even finished most of the research for a novel about the gold fever in California. I would sit before my computer with my head filled with ideas but be unable to transfer them to the screen. "You have to give yourself time, you're still grieving," my mother reminded me in her letters, and Abuela Hilda softly repeated the same advice. During that time, she was taking turns between staying at her daughter's house in Chile and then ours or Nico's in California. This kind woman, the mother of Hildita, my brother Pancho's first wife, had been warmly adopted as grandmother by all of us, especially Nico and you, whom she spoiled from the moment you were born. She was my accomplice in any madness I dreamed up in my youth and companion in your and Nico's adventures.

Marijuana and Silicone

Abuela Hilda, tireless, tiny, and cheerful, had managed through a lifetime to avoid things that might cause her anguish. That was probably the secret of her astounding disposition. She had the mouth of a saint: she never spoke ill of anyone, she fled from arguments, she quietly tolerated others' stupidity, and she could make herself invisible at will. One time when she had a full-blown case of pneumonia, she kept on her feet for two weeks, until her teeth began to chatter and fever misted her eyeglasses; only then did we realize that she was near leaving us for the other world. She spent ten days in an American hospital where no one spoke Spanish, mute with fright, but if we asked how she felt, she said she was doing fine, and added that the Jell-O and yogurt there were better than what we had in Chile. She lived in a fog; she didn't speak English, and we would forget to translate the medley of tongues we spoke in our house. Since she couldn't understand the words, she observed body language. A year later, when Celia's drama erupted, she was the first to have her suspicions, since she picked up signals that the rest of us simply didn't notice. The only medications she took were some mysterious green pills she tossed into her mouth when the atmosphere around her grew too tense. She could not deny your death, Paula, but she pretended that you were on a trip and spoke of you in the future tense, as if she would see you the next day. She had limitless patience with my grandchildren, and although she weighed ninety-five pounds and had the bones of a turtledove, she carried Nicole everywhere she went. We were afraid my youngest grandchild would be fifteen years old before she learned to walk.

"Cheer up, mother-in-law! What you need for literary inspiration is a joint," Celia counseled. She had never smoked marijuana, but she was dying to try.

"Pot clouds your mind, it won't inspire you at all." Tabra's opinion came from experience. She was on her way back from such experiments.

"Why don't we try it?" asked Abuela Hilda, to put an end to our doubts.

And that was how the women of the family ended up smoking marijuana at Tabra's house, having told everyone that we were going on a spiritual retreat.

The evening began badly, because Abuela wanted Tabra to pierce her ears and the ear-piercing gun jammed and stuck in her earlobe. When Tabra saw the blood, her knees buckled, but Abuela did not lose her composure. She held the apparatus, which weighed over a pound, until Nico arrived an hour later, equipped with his toolbox. He dismantled the gun, and freed her. The bloody ear had doubled in size. "Now, Tabra, pierce the other one for me," Abuela requested. Nico stayed long enough to take the gun apart a second time, and then left out of respect for our "spiritual retreat."

During the process of the ear piercing, Tabra's breasts several times brushed against Abuela Hilda, who kept looking at them out of the corner of her eye, until finally she couldn't stand it any longer and asked what it was she had in them. My friend speaks Spanish, so she was able to explain that it was silicone. She told Abuela that when she was a young schoolteacher in Costa Rica, she had a rash on her arm and had to go to the doctor. He asked her to take off her blouse, and when she explained that the problem was confined to one arm, he insisted. She took it off. "Woman!

You're flat as a pancake!" he exclaimed when he saw her. Tabra realized it was true, and then he suggested a solution that would benefit them both. "I intend to specialize in plastic surgery but I don't have patients yet. What do you think about letting me experiment on you? I won't charge you anything for the operation, and I'll give you some knockout tits." It was such a generous proposition, and expressed so delicately, that Tabra couldn't refuse. Nor did she dare refuse when he showed a certain interest in going to bed with her, an honor accorded to only a few of his patients, the doctor made clear. She did, however, refuse when he wanted to extend his offer to her younger sister, who was only fifteen. And that was how Tabra had ended up with her marble prostheses.

"I've never seen such hard boobs," commented Abuela Hilda.

Celia and I had to touch them too, and then we wanted to see them. No question about it, they were strange; they looked like footballs.

"How long have you been carrying this burden around, Tabra?" I asked.

"Oh, about twenty years."

"Someone needs to examine you, this doesn't seem normal."

"Don't you like them?"

The rest of us women took off our blouses to compare. Ours would never be spread across the pages of men's magazines, but at least they were soft to the touch, as nature created them, and not like hers, which had the consistency of truck tires. My friend agreed to let us take her to see a specialist, and soon after there began what we in the family called "the odyssey of the boobs," a series of unfortunate mishaps, the setting of which was the office of a plastic surgeon and the one advantage the fact that it solidified my friendship with Tabra.

At nightfall we built a bonfire among the trees and roasted hot dogs and toasted marshmallows on sticks. Then we lit one of the joints, which had cost us no little trouble to obtain. Tabra inhaled a couple of times, announced that pot made her meditative, closed her eyes, and dropped like a stone, anesthetized. We carried her back to the house, no small job, deposited her on the floor, covered her with a throw, and we went back to the shelter of the blossoming trees in the garden. There was a full moon, and the stream, fed by rain, was leaping among the stones in its bed. Celia played her guitar and sang her most nostalgic songs, and Abuela started knitting between joints, which had not had the effect of making us high, as we'd expected, but produced only giggles and insomnia. We stayed in Tabra's woods, telling each other stories about our lives until dawn, when Abuela announced that it was time to have a whisky, seeing that the marijuana hadn't even warmed her bones. Ten hours later, when Tabra came to and checked the ashtray, she calculated that we'd smoked a dozen joints with no visible consequences, and deduced, to her amazement, that we were invulnerable. Abuela thought the cigarettes must have been filled with straw.

The Angel of Death

In early autumn, when an unusual climate of peace was reigning in the house and we were beginning to abandon ourselves to a dangerous complacency, we received a visit from an angel of death. It was Jennifer's companion, with his dark, puffy face and the dull eyes of a hard drinker. In his drawn-out, unintelligible jargon, which Willie could barely decipher, he told us that Jennifer had disappeared. He hadn't heard anything from her since she'd left three weeks before to visit an aunt in another city. According to the aunt, the last time she had seen her, she'd been in the company of some rough-looking men who'd come by in a van and picked her up. Willie reminded the man that often months went by without news of his daughter, but he repeated that she had disappeared, and added that she had been very sick and could not have gone far in her condition. Willie began a systematic search through jails and hospitals; he spoke with the police, contacted federal agencies in case his daughter had crossed a state line, and hired a private detective, all without success. In the meantime, Fu and Grace set the members of the Zen Center, and I my Sisters of Disorder, to praying. The story the man told us smelled fishy to me, but Willie assured me that in such cases the first person to be suspected in the eyes of the law is the one living with the missing person, especially if he has a long rap sheet, as this man did. No question that he had been thoroughly investigated.

We're told that there is no pain as great as that of the death of a child, but I believe it has to be worse when your child disappears and you live forever not knowing what happened to her. Did she die? Did she suffer? You keep the hope that she is alive, but constantly wonder what kind of life she's living and why she doesn't get in touch with her family. Every time the telephone rang late at night, Willie's heart stopped with hope and with terror. It could be Jennifer's voice, asking him to come look for her somewhere, but it might also be the voice of a policeman asking him to come to the morgue to identify a body.

Months later, Jennifer still had not turned up, but Willie clung to the idea that she was alive. I don't know who it was who suggested that he should consult a psychic who sometimes helped the police solve cases. She had the gift of locating bodies and missing persons, and that was how we ended up together in the kitchen of a dilapidated house near the port. The psychic looked nothing like a divine, no star-patterned skirts, kohl-rimmed eyes, or crystal ball. She was a fat woman in tennis shoes and an apron, who kept us waiting while she finished bathing her dog. In the kitchen—narrow, clean, and orderly—were a pair of yellow plastic chairs, and we took our seat in them. Once the dog was dry, she offered us coffee, and sat down on a small stool facing us. We sipped from our mugs in silence a few minutes, then Willie explained the reason for our visit and showed her a series of photos of his daughter: some in which she was more or less healthy, and the most recent, taken in the hospital, already very ill, with Sabrina in her arms. The psychic examined them one by one, then put them on the table, placed her hands on them, and closed her eyes for long minutes. "Some men took her in a vehicle," she said finally. "They killed her. They dropped the body in a woods near the Russian River. I see water and a wood tower; it must be a ranger's lookout."

Willie, pale as death, said nothing. I put the payment for her services on the table, three times what it costs to see a physician, took my husband by the arm, and pulled him to the car. I got the key from his pocket, pushed him into the passenger side, and I drove, hands trembling and eyes clouded, across the bridge toward home. "You shouldn't believe any of this, Willie. It isn't science, she's a quack," I begged him. "I know that," he replied, but the harm had been done. Even so, he didn't truly grieve until much later, when we went to see a film about the death penalty, *Dead Man Walking,* in which there is a scene of the murder of a girl in a forest, similar to what the psychic had described. In the silence and darkness of the theater, I heard a heart-rending cry, like the howl of a wounded animal. It was Willie, doubled over in his seat, with his head on his knees. We felt our way out of the theater, and once in the parking lot, locked in our car, he wailed for his lost daughter.

One year later Fu and Grace offered to have a ceremony in the Zen Center in Jennifer's memory, to give dignity to that tragic life and closure to the obscure death that left the family in eternal suspense. Our small tribe, and a few friends, including Tabra, Jason, Sally, and Jennifer's mother with a few of her friends, met in the same room where we had celebrated Sabrina's first birthday, in front of an altar that held pictures of Jennifer in her best days, flowers, incense, and candles.

They had placed a pair of shoes in the center of the circle to represent the new path she had taken. Jason and Willie were moved by the good intentions of all those present, but they couldn't avoid exchanging smiles because Jennifer had never had a pair of shoes like those on her feet; they should have found some purple sandals, something more appropriate to her style. Both of them, who knew her well, imagined that if she had been watching that sad reunion from above she would be rolling with laughter, for she thought that anything with a hint of New Age was ridiculous, and besides she wasn't a person to mourn. She was completely lacking in self-pity; she was daring and bold. Without the addictions that trapped her in a life of misery, she might have lived an adventurous life, because she had her father's strength. Of Willie's three children, only Jennifer inherited Willie's lion heart, and she passed it on to her daughter. Sabrina, like Willie, can be dropped to her knees, but she always gets back up on her feet. That little girl, who almost never even saw her mother but who had her image engraved on her soul before she was born, participated in the rites in Grace's arms. At the end, Fu gave Jennifer a Buddhist name, U Ka Dai Shin: wings of fire, great heart. It was a proper name for her.

In a period of quiet meditation during the ceremony, Jason thought he heard his sister's voice breathing into his ear. "What the fuck are they doing? They don't have the least idea what happened to me! For all they know I could still be alive, right? The joke is that they'll never know." Maybe for that reason, Jason has never stopped looking for her, and now, all these years later, now that we have the DNA tests, he is stubbornly trying to locate her in the infinite police archives of unsolved tragedies. As for me, during the meditation a scene emerged with great clarity in my mind: Jennifer was sitting on the bank of a river, paddling her feet and tossing little stones into the water. She was wearing a summer dress and she looked young and healthy, with no trace of pain. Rays of sun shone through the leaves and illuminated her blond hair and slim body. Suddenly she lay down, curled up on the mossy ground, and closed her eyes. That night I told Willie about my vision and we decided that that was how Jenny had died, and not the way the psychic had told us. She was very tired, she slept, and she never woke up. The next morning we got up early and the two of us went to the forest. We wrote Jennifer's name on a piece of paper, burned it, and threw the ashes into the same stream where earlier we had scattered yours. You two didn't know each other in this world, Paula, but we like to imagine that maybe your spirits are playing among the trees like sisters.

Family Life

In the spring of 1994 Rwanda was often in the newspapers. News of the genocide was so horrible that it
was difficult to believe. Children were being hacked to
death, pregnant women were ripped open with knives
to tear the fetuses from their wombs, entire families
were murdered, hundreds of starving orphans were
wandering the roads, villages were burned with all
their inhabitants.

"What does the rest of the world care about what's
happening in Africa? It's only blacks that are dying,"
Celia commented indignantly, with that incendiary
passion she dedicated to nearly any subject.

"The killing in Rwanda is terrible, Celia, but I don't
think that's the only reason you're depressed. Tell me
what's really happening with you," I prodded.

"Imagine if they were hacking my children apart!" and she burst out crying.

There was no doubt that something was brewing in the soul of my daughter-in-law. She didn't have a moment of peace; she ran around doing a thousand tasks, she hid around corners to cry, and every day she was more emaciated. In addition, she had developed a true obsession with bad news, which she discussed with Jason, the only one in the family who read all the newspapers and who was capable of analyzing events with a journalist's instinct. He was the first person I heard relate religion with terror, long before fundamentalism and terrorism were practically synonymous. He explained to us that the violence in Bosnia, the Middle East, and Africa, the excesses of the Taliban in Afghanistan, and other disconnected events were caused by religious as well as racial hatreds.

Jason and Sally were talking about moving as soon as they could find an apartment somewhere, but they had looked in vain to find something within range of their limited income. We offered to help them, without insisting too much, so we wouldn't give them the impression that we were throwing them out. It was pleasant to have them with us. I was moved to see Jason in love for the first time and talking about getting married, though Willie was convinced that Sally and his son would not make a good pair. I don't know where he got that idea, they seemed to get along very well.

Abuela Hilda stayed in California for long periods of time, and under her influence the house would become a gambling den. Even my grandchildren, innocents still sucking their pacifiers, learned to do tricks with cards. She showed them how to play with such skill that a few years later Alejandro could have earned a living with a deck of cards. Once, when he was a runny-nosed kid of ten, a little sandpiper of a boy with round eyeglasses and teeth like a beaver, he wandered into a group of ominous-looking types who had brought their trailers and motorcycles to the beach and set up camp. The general look of those men—tank tops, tattoos, cowboy boots, and the inevitable bellies of good beer drinkers— did not intimidate Alejandro because he saw that they were playing cards. He went over to them, totally self-assured, and asked if he could sit in. They answered with loud guffaws, but he insisted. "We play for money here, kid," they warned him. Alejandro nodded, feeling rich because he had five dollars in small change. They told him to sit down, and offered him a beer, which he amiably refused, more interested in the game. After twenty minutes, my grandson had fleeced the seven tough guys and left with bills spilling out of his pockets, followed by a hailstorm of oaths and curses.

We lived as a tribe, Chilean style; we were almost always together. Abuela had such a good time with Celia and her children that she preferred her company a thousand times more over mine and spent long stretches in her home. We had explained to Abuela that Sabrina's mothers were lesbians, Buddhists, and vegetarians, three words she was not familiar with. The vegetarianism was the only thing that seemed unacceptable to her, but she made friends with them anyway. More than once she visited them at the Zen Center, where she induced them to eat hamburgers, drink margaritas, and bet on poker. At times my mother and Tío Ramón, my ineffable stepfather, would come from Chile, and, added to them, my brother Juan, who arrived from Atlanta with the tilted head and grave expression of a bishop: he was studying theology. After he had devoted four years to the divine, Juan graduated with honors and then decided he wasn't cut out to be a preacher and went back to his university position as professor of political science, where he is today. Willie bought food wholesale and cooked for that camp of refugees. I see him in the kitchen, bloody knives attacking a hindquarter of beef, frying bags of potatoes, and chopping tons of lettuce. In moments of inspiration he would make lethally hot Mexican tacos while listening to his *ranchera* CDs. The kitchen would look like the morning after a night of Carnival, and guests would lick their chops but later pay the consequences of an excess of grease and chilis.

Our house was magical; it stretched and shrank according to the need. Perched halfway up a hill, it offered a panoramic view of the bay; there were four bedrooms on the first floor and an apartment below. It was there in 1992 that we installed a hospital room where you spent several months without altering the rhythm of family life. Some nights I would wake to the murmur of my own memories and those of the characters escaped from other people's dreams. I would get up and roam through the rooms, grateful for the quiet and warmth of that house. Nothing bad can happen here, I would think, all the evil has been expelled, and Paula's spirit is looking after us. Sometimes the dawn would surprise me with its capricious colors of watermelon and peach. I liked to look down at the scene at the foot of the hill, with the fog rising from the lagoon and wild geese flying south.

Celia was just recovering from the battering of her three pregnancies when she had to go to Venezuela for her sister's wedding. By then she had a residence visa that allowed her to travel abroad and return to the United States. Nico and the children temporarily moved over to our house, a solution that Abuela found ideal: "Why don't we all live together, the way we should?" she asked. In Venezuela Celia was confronted by everything she had wanted to leave behind when she married Nico. It can't have been pleasant because she returned with her spirits lower than low, having decided to sever all contact with most of her relatives. She clung to me, and I was prepared to defend her against anything that came along, even herself. She started losing weight again, but we had a family council and forced her to see a specialist, who prescribed therapy and anti-depressants. "I don't believe in any of this," she told me, but the treatment helped, and soon she was playing the guitar and making us laugh and rant with her antics. Despite the inexplicable fits of sadness, she had blossomed with maternity.

Her children were a perpetual circus, and Abuela often reminded us that we must cherish them because they soon grow up and leave home. It was the children, more than the doctors' prescriptions, that kept Celia going during that time. Alejandro, a little shy but very smart, stuttered wise phrases with his mother's deep voice. That year at Easter, before he went outside with his basket to hunt for Easter eggs in the garden, he whispered to me that rabbits don't lay eggs because they're mammals. "Then who leaves the Easter eggs?"

I asked, like a fool. "You," he answered. Ever since Nicole, the youngest, could stand, she'd had to defend herself against her brother and sister. One birthday I had the bad idea of giving Alejandro—who had begged on his knees, batting his giraffe eyelashes—a game of plastic Ninja daggers. First I got specific authorization from his parents, who did not allow him to have weapons, just as they opposed television—both impractical California New Age ideas because you can't raise little ones in a bubble. Better for them to be contaminated while they're young, that's the best way to immunize them. I warned my grandson that he could not attack his sisters, but that was like handing him candy and telling him not to suck it. He had the daggers all of five minutes before he stabbed Andrea, who gave it right back to him, and then both of them turned on Nicole. Next we saw Alejandro and Andrea running for their lives, with Nicole close behind, a dagger in each hand, howling like a serial thriller Apache. She was still in diapers. Andrea was the most colorful. She always wore pink; except for a pair of lime-green plastic sandals; her golden curls peeked from among the adornments she put in her hair—tiaras, ribbons from packages, paper flowers—and she lived lost in her imaginary world. She also had her "pink power," a magic ring with a stone of that color, a gift from Tabra, which could convert broccoli into strawberry ice cream and send a kick by long distance to the boy who had made fun of her at recess. Once Andrea's teacher raised her voice to her, and my granddaughter confronted her squarely, pointing the finger with the powerful ring. "Don't you dare speak to me that way. I am *Andrea!*" Another time she came back from school very upset, and hugged me.

"I've had a miserable day," she confessed, sobbing.

"Wasn't there even one good minute, Andrea?"

"Yes. A girl fell and broke two teeth."

"But my God, Andrea, what's good about that!"

"It wasn't me."

Messages

Paula was published in Spain with a cover photo Willie had taken. In it you are smiling and full of life, and your dark hair is covering you like a mantle. Soon hundreds of letters began pouring in; we had to store them in boxes in the office and Celia couldn't find enough time to sort and answer them. For years I had received letters from enthusiastic readers, although I admit that not all of them were motivated by appreciation for my books. Some were requests, like the one from the author of sixteen unpublished novels who gallantly offered to work with me and share royalties fifty-fifty, and another came from a couple of Chileans in Sweden who asked me to buy them tickets back to Chile; after all, it was my uncle Salvador Allende's fault that they'd had to leave the country. Even so, nothing could compare with the avalanche of correspondence that flooded in with the publication of *Paula*. I tried to answer every one, even if with only a couple of lines scrawled across a card, because each letter had been written from the heart and sent out blind, some to my publishers, others to my agent, and many through friends and bookstores. I spent part of the night making cards with the Japanese papers Miki Shima gave me and little pieces of silver and semiprecious stones from Tabra. The letters that came were so heartfelt that years later, when the book had been translated into several languages, some European publishers decided to publish a selection of that correspondence. Sometimes parents who had lost children wrote me, but most were young people who identified with you, including girls who wanted to meet Ernesto, in love with the widower without knowing him. Tall, well built, dark, and tragic, he attracted women. I don't think he wanted for consolation; he isn't a saint, and celibacy isn't his forte, as he himself has told me, and as you always knew. Ernesto always swears that if he hadn't fallen in love with you, he would have entered seminary and become a priest, but I doubt it. He needs a woman at his side.

Occupied with the letters, I had no time for writing, and even my exchange with my mother slowed down. Instead of the daily messages that had kept us united for decades, we talked by phone or sent brief faxes, avoiding confidences that might be exposed to a stranger's curiosity. Our correspondence during that period is very boring. Nothing like the mail, the good old snail mail with its privacy. Nothing like the pleasure of waiting for the mailman, opening the envelope, taking out the pages my mother had folded, and reading the then two-weeks-old news. If it was bad, it didn't matter any longer, and if it was good, it was never too late to celebrate.

Among the letters came one from a young nurse who had attended you in the intensive care unit in the hospital in Madrid. Celia was the one who saw it first. She brought it to me, pale as wax, and we read it together. The nurse said that after reading the book she had felt as if it was her duty to tell me what had happened. Medical incompetence and a power failure that interrupted the oxygen had caused severe brain damage. Many people knew what had happened but tried to hide it, perhaps with the hope that you would die and they would avoid an investigation. For months, the nurses had watched me waiting all day long in the corridor of lost steps, and they had often wanted to tell me the truth but didn't dare face the consequences. The letter left me reeling for days. "Don't think about it, daughter, there's nothing we can do about it now," my mother wrote when I told her. "That was Paula's fate. Now her spirit is free. Your daughter will never have to suffer the troubles life always visits upon us." Right. And following that reasoning, we would all be better off dead, I thought.

That memoir brought more interest from the public and the press than all my previous books combined. I made lots of trips, gave hundreds of interviews, dozens of readings and lectures, and signed thousands of books. One woman wanted me to inscribe nine books for her, one for each of her friends who had lost a child, and one for her. Her daughter had been left a paraplegic in an automobile accident, and as soon as she could manage a wheelchair, she drove it into a swimming pool. Pain and more pain. By comparison, mine was bearable; at least I had been able to take care of you to the end.

Four Minutes of Fame

The movie based on my first novel, *The House of the Spirits*, was announced with great fanfare because it had a formidable cast of the great stars of the day: Meryl Streep, Jeremy Irons, Glenn Close, Vanessa Redgrave, Winona Ryder, and, my favorite, Antonio Banderas. Now, several years later, when I think of them, those actors seem as far away in time as the stars of the silent screen. Time is implacable. When my first novel was published, several members of my mother's family were upset with me, some because our political ideas are diametrically opposed, and some who believed I had betrayed family secrets. "Dirty linen is washed at home" is Chile's watchword. To write that book I had used my grandparents, some uncles, and other bizarre characters in my large Chilean tribe as models, as well as political happenings of the time and anecdotes I'd listened to my grandfather tell for years, but I'd never imagined that some people would take it literally. Mine is a twisted and exaggerated version of events. My grandmother could never move a billiard table with the power of her mind, like Clara del Valle, nor was my grandfather a rapist and murderer, like Esteban Trueba in the novel. For many years those relatives didn't speak to me, or at best avoided me. I thought that the film would be like throwing salt in the wound, but it was just the opposite. The power of film is so enormous that the movie became the official history of my family, and I have found that now photographs of Meryl Streep and Jeremy Irons have replaced those of my grandparents. It was rumored that the film would sweep the Academy Awards in Hollywood, but before it was even shown negative articles began to appear recounting how Hispanic actors had not been given the chance to work in a film with a Latin American theme. In the early days, they wrote, when a black actor was needed on the screen they painted a white with shoe polish, and now that a Latino was required, they pasted a mustache on a white man. The movie was filmed in Europe by a Danish director, with German money, Anglo-Saxon actors, and an English sound track. There was little about it that was Chilean, but to me it seemed better than the book and I was sorry that bad will was building in advance. Months earlier, the director, Bille August, had invited Willie and me to the filming in Copenhagen. The exteriors were done on an estate in Portugal, which later became a tourist attraction, and the interiors in a house built on a studio set in Denmark. The furniture and decor were rented from antique stores in London. I recall that there was a little enamel box I wanted to slip into in my pocket as a souvenir, but each object had a coded number, and someone was charged with keeping track of them. Then I asked for the head of Vanessa Redgrave, but they didn't give it to me. I'm referring to a wax model that was supposed to appear in a scene in a hat shop but was omitted for fear of causing hilarity in the audience rather than the desired fright. I wonder what ever became of that head? Perhaps Vanessa has it on her night table, to remind her of just how tenuous life is. I would have used it for years to break the ice in conversations, and to scare my grandchildren. In the cellar I kept hidden skulls, pirate maps, and treasure trunks; nothing better than a childhood of terror to stimulate the imagination.

For a week, Willie and I rubbed elbows with celebrities, and lived as important people in this world live. Each star had a court of assistants, makeup artists, hairdressers, masseuses, and cooks. Meryl Streep, beautiful and remote, came and went with her children and their respective nannies and tutors. One of her young daughters with the same talent and ethereal appearance as her mother acted in the film. Glenn Close had several dogs, and assistants to look after the dogs. She had read my book very carefully to prepare for the part of Férula, the spinster, and we spent pleasant hours chatting. She asked me whether the relationship between Férula and Clara was lesbian, and I didn't know what to answer as I was surprised by that reading. I think that in the early years of the twentieth century in Chile, a time in which part of the novel is set, there were loving relations among women that never reached a sexual level because of the societal and religious impediments of the time. In real life Jeremy Irons was not precisely the frosty English aristocrat we admire on the screen; he could have been a likable taxi driver in the suburbs of London. He had a dark sense of humor, fingers stained with nicotine, and was proud of his inexhaustible repertoire of wild stories, such as one in which he loses his dog in the London Underground, and for one entire morning dog and master cross paths going in opposite directions, leaping out of trains every time they catch sight of each other in some station. I don't know why, but for the film they put something like a bit in his mouth that distorted his face and his voice. Vanessa Redgrave, tall, patrician, luminous, with eyes of cobalt blue, showed up sans makeup and with a babushka around her head, none of which diminished one whit the formidable impact of her presence. Winona Ryder I met later; she was a kind of pretty little boy whose mother had whacked his hair with her scissors; to me she seemed enchanting, but among the technical crew she had a reputation for being spoiled and capricious. I've heard that later that damaged a career that could have been brilliant. As for Antonio Banderas, I had seen him once or twice before and was already in love with him, one of those shy, ridiculous schoolgirl crushes adolescents have on screen stars—no matter that if you stretched the years a little he could be my son. There was always a line of half-frozen fans at the main door of the hotel, feet buried in the snow, hoping some star would come by and they could ask for an autograph, but the actors all used a service door, and fans had to be content with my signature. "Who's she?" I heard one ask in English, pointing to me. "Can't you see? She's Meryl Streep," another answered.

Just when we'd become accustomed to living like royalty, our vacation ended. We went home and immediately we passed into absolute anonymity; if we called any of our famous "friends," we had to spell our names. The world premiere wasn't held in Hollywood but in Munich—the producers were German—where we were greeted by a throng of tall people and a crushing bombardment of cameras and lights. Everyone was wearing black, and I, in the same color but only half their height, disappeared below belt-buckle level. I appear in only one press photo, and I look like a terrified mouse, black on black, with Willie's amputated hand on one shoulder.

I am going to tell you something now that happened a long time after the movie version of *The House of the Spirits* came out, or I'll never tell it at all; it is in reference to fame and that's something that never interested you, Paula. I was asked to carry the Olympic flag in the Winter Games in Italy, in February of 2006. It took only four minutes for me to be catapulted into fame. Now people recognize me in the street, and at last my grandchildren boast of having me for a grandmother.

One day Nicoletta Pavarotti called me; she is the wife of the tenor, a charming woman thirty-four years younger than her famous husband, and she told me that I had been selected as one of the eight persons who would carry a flag in the opening ceremony of the Olympic Games. I replied that there was surely some error because I am the opposite of an athlete, and in truth, I wasn't sure I could make it around the stadium without a walker. She explained that it was a great honor; the candidates had been rigorously screened, and their lives, their ideas, and their work had been thoroughly investigated. In addition, it would be the first time that only women would carry the flag: three female gold medalists and five women representing continents. I represented Latin America. My first question, naturally, was what I would wear. She replied that we would all be dressed in a uniform, and she asked for my measurements, which sent me into a panic: I could see myself in a quilted outfit in some repulsive pastel color, fat as an ad for Michelin tires.

"May I wear high heels?" I asked, and heard a sigh from the other end of the line.

In mid-February Willie and I, with the rest of the family, arrived in Turin, a beautiful city on an international scale, but not to Italians, who are not impressed even by Venice or Florence. Enthusiastic crowds cheered as the Olympic torch was carried through the streets, or as any of eighty competing teams passed by, each in their national colors. Those young people were the best athletes in the world; they had trained since they were three or four years old and had sacrificed everything to participate in the Olympiad. They all deserved to win, but there is an unpredictable element of luck involved: one flake of snow, one centimeter of ice or the wind speed can determine the result of a race. Nonetheless, what weighs heaviest, more than training or luck, is heart: only the most valiant and determined heart will take home the gold medal. Passion, that is the winner's secret. The streets of Turin were covered with posters proclaiming the motto of the Olympiad: "Passion Lives Here." And that is my greatest wish, to live passionately to the very end.

In the stadium, I met the other flag bearers: three athletes and the actresses Susan Sarandon and Sophia Loren, as well as two activists, the Nobel Peace laureate Wangari Maathai from Kenya and Somaly Mam, who campaigns against the sexual traffic of children in Cambodia. I was also given my uniform. It wasn't the style of clothing I normally wear but it wasn't as horrible as I had imagined: sweater, skirt, and coat of winter white wool, boots and gloves of the same color, all bearing an expensive designer's logo. Not that bad, to tell the truth. I looked like a refrigerator, but the others did, too, except for Sophia Loren, tall, imposing, full-breasted, and sensual, and splendid in all her seventy-some years. I don't know how she keeps herself slim, because during the long hours we were waiting in the wings, she never stopped snacking on carbohydrates: cookies, nuts, bananas, chocolates. And I don't know how she can be so tanned and not have wrinkles. Sophia is from another era, very different from today's models and actresses, who look like skeletons with false breasts. Her beauty is legendary, and apparently indestructible. Earlier I had heard her say during a television program that the secret for beauty was to maintain good posture and not "make old woman noises," no moaning, grunting, coughing, puffing, talking to yourself, or breaking wind, though you have nothing to worry about, daughter, you will always be twenty-eight years old. On the other hand I, who am hopelessly vain, have tried to follow that advice in every detail, since I cannot imitate Sophia in any other way.

The woman who impressed me most was undoubtedly Wangari Maathai, who works with women in African villages and has planted more than thirty million trees, changing the soil and the weather in some regions. This magnificent woman glows like a lamp, and I felt an irresistible urge to put my arms around her, something that occasionally happens in the presence of certain young men, but never with a lady like her. I hugged her for a long time, unable to let go, she was like a tree: strong, solid, quiet, content. Wangari must have been frightened, and she unobtrusively pushed me away.

The Olympic Games opened with an extravagant spectacle in which thousands of actors, dancers, extras, musicians, technicians, producers, and many more people took part. At a certain moment, about eleven p.m., when the temperature had fallen to below zero, we were led to the wings and given the enormous Olympic flag. Loudspeakers announced the climactic moment of the ceremony and the Triumphal March from Verdi's *Aïda* was rousingly sung by the 40,000 spectators. Sophia Loren walked ahead of me. She is a head taller than I am, not counting her luxuriant hair, and she walked with the elegance of a giraffe on the African savannah, supporting the flag on her shoulder. I trotted behind on tiptoe, holding my section above my head with my arm extended. I was dwarfed beneath the damned flag. Of course all cameras were focused on Sophia Loren, the eternal symbol of beauty and sexuality, and that worked in my favor because I appeared in all the press photographs, even though between Sophia's legs. I confess to being so happy that according to Nico and Willie, who with tears of pride were urging me on from the gallery, I was levitating. That circuit of the Olympic stadium was my four magnificent minutes of fame. I have kept all the articles and photos because it is the one thing I do not want ever to forget when senile dementia erases all my other memories.

The Depraved Santa Claus

B ut let's go back, Paula, and not get lost in time. We grew very fond of Sally, Jason's sweetheart, a discreet girl of few words who kept herself in the background, although she was always attentive and participating. She had a fairy godmother touch with the children. She was short, pretty without being flamboyant, with smooth blond hair and never a drop of makeup. She looked about fifteen. She had a job in a center for juvenile delinquents, which required courage and a strong hand. She got up early, left, and we wouldn't see her until evening, when she came home dragging with fatigue. Several of the youths in her charge had been arrested for assault with a deadly weapon, and although they were minors, they were the size of mastodons. I don't know how—beside them she looked like a sparrow—she earned their respect. The day one of the troublemakers threatened her with a knife was the day I offered her a safer job in my office helping Celia, who by now could not keep up with the load of work. They were very good friends; Sally was always ready to help with Celia's children and spend time with her because Nico was gone, working and studying English. Over time, I came to know Sally, and I agreed with Willie that she had very little in common with Jason. "Keep your nose out of it," Willie ordered. But how could I do that? They lived in our house, and Sally's bridal gown, a vision of meringue-colored lace, was hanging in my closet. She and Jason planned to get married as soon as he finished his studies. That was what Jason told us, but Sally showed no sign of impatience; they acted like a pair of bored fifty-year-olds. These modern courtships, long and easygoing, worry me. Urgency is inseparable from love. According to Abuela Hilda, who saw things that were invisible, if Sally married Jason it would not be because she loved him, but to stay in our family.

The only work Jason could find after graduating from college was a temporary job in a mall, sweating in a ridiculous Santa Claus suit. At least it had the effect of teaching him that he would have to continue his education and get a professional degree. He told us that most of the Santa Clauses were poor devils who came to work with several jolts of cheap whisky under their belts, and that some fondled the children. In view of those revelations, Willie decided that our children would have their own Santa Claus, and he bought a convincing beard, patent leather boots, and a splendid costume of red velvet trimmed with real rabbit fur. I wanted him to choose something less expensive, but he proclaimed that he never wore anything ordinary, and besides, it would serve many years and the cost would be amortized. So that Christmas we invited a dozen children, with their parents, and at the appointed hour we turned down the lights, someone played Christmas music on an electronic organ, and Willie came in through a window, carrying his bag of gifts. His entrance produced a stampede of terror among the youngest, except for Sabrina, who is not afraid of anything. "You must be very rich if you can get Santa on such a busy night," she commented. The older children were enchanted, until one of them declared that he didn't believe in Santa Claus, and Willie angrily replied, "Then no presents for *you*, you little shit!" That was the end of the party. The children immediately suspected that it was Willie hiding behind the beard—who else would it be?—but Alejandro put an end to any speculation with this irrefutable logic. "We don't want to know. It's like the tooth fairy that brings money when you lose a tooth. It's best if our parents think we're stupid." That year Nicole was still too young to participate in the farce, but three years later she was consumed with doubt. She was terrified of Santa, and every Christmas we had to stay in the bathroom with her, where she closed the door and shivered until we assured her the terrible old man had left on his sleigh for the next house. This time she hunkered down beside the toilet, wearing a long face and refusing to open her presents.

"What *is* the matter, Nicole?" I asked.

"Tell me the truth. Is Willie Santa Claus?"

"I think it would be better if you asked him," I suggested, afraid that if I lied to her, she would never believe me again.

Willie led her by the hand to the room where he kept the costume he had just worn, and admitted the truth. He cautioned her that this would be a secret between the two of them, one she shouldn't share with the other children. My youngest grandchild returned to the party with the same long face, took her place in a corner, and wouldn't touch her presents.

"And what's the matter now, Nicole?"

"You've always made fun of me! You've ruined my life!" was her answer. She was not yet three years old.

I told Jason how helpful my training as a journalist had been in my work as a writer, and suggested that it could be a first step toward his literary career. Journalism teaches you to investigate, sum up, work under pressure, and use language efficiently, and in addition forces you to keep the reader in mind, something authors tend to forget when preoccupied with posterity. After a lot of pressure—he doubted himself and didn't even want to fill out the admission forms—he applied to several universities and to his surprise was accepted in all of them. He could give himself the pleasure of studying journalism in the most prestigious of them all, Columbia University in New York. That put physical distance between him and Sally, and it seemed to me that their lukewarm relationship would become frigid, although they kept talking about getting married. Sally stayed close with us, working with Celia and me and helping with the children. She was the perfect aunt.

Jason left in 1995 with the idea of graduating and returning to California. Of all Willie's children, he was the one who most liked the idea of living in a tribe. "I want to have a big family, and this blend of Americans and Latins works great," he told me once. To fit in, he had spent a few months in Mexico studying Spanish, and spoke it well, with the same bandit accent as Willie's. Jason and I were always friends; we shared the vice of books, and we liked to sit on the terrace with a glass of wine and tell each other plots for possible novels. He felt that you, Ernesto, Celia, and Nico were as much his siblings as the ones fate had given him, and he wanted all of us to be together forever. However, after your death and Jennifer's disappearance, we all sank into sadness, and bonds were cut or altered. Jason says now, years later, that the family went to hell, but I remind him that families, like almost everything in this world, metamorphose or evolve.

An Enormous Rock

C elia and Willie argued at the top of their lungs, as passionate over trivialities as they were about serious matters.

"Put on your seat belt, Celia," he would say to her in the car.

"You don't have to in the back seat."

"Yes, you do."

"No!"

"I don't give a shit whether you have to or not! This is my car and I'm driving. Put it on or get out!" Willie would roar, red with rage.

"Fuck you, I'm getting out!"

Celia had rebelled against masculine authority from the time she was a child, and Willie, who himself exploded at the least provocation, accused her of being an ill-mannered little brat. He was often furious with her, but everything was forgiven as soon as she took up her guitar. Nico and I tried to keep them separated, but we were not always successful. Abuela Hilda stayed out of it; the most she ever said was that Celia was not accustomed to accepting affection, but that with time she'd get the drift of it.

Tabra was operated on to remove the footballs and replace them with normal breasts, sacks of a solution less toxic than the silicone. As an aside, the doctor who put in the original ones has become one of the most famous plastic surgeons in Costa Rica, so the experience he gained with my friend was not wasted. I suppose that by now he must be a doddering old man and doesn't even remember the young American girl who was his first experiment. Tabra this time was six hours in the operating room. They had to scrape the fossilized silicone off her ribs, and when she came out of the clinic she was in such pain that we brought her home to stay until she could get along by herself. Her lymph nodes were so inflamed she couldn't move her arms, and she had a reaction to the anesthesia that left her nauseated for a week. She couldn't keep anything down except watery soups and toast. By coincidence, Jason had left for New York to study and Sally had moved to an apartment she shared with a friend, but Abuela

Hilda, Nico, Celia, and the three children were temporarily living with us. The Sausalito garret had become too small for Nico's family and we were in the final stages of buying a house for them; it was a little farther away and needed work but it had a pool, a lot more room, and untouched wooded hills at the back door, perfect for bringing up the children. Our house was filled to the brim, but in spite of how bad Tabra felt, the atmosphere was usually festive, except when Celia or Willie got heated about something and then the least spark set off a fight. The day Tabra arrived, one exploded over something relatively serious that had happened in the office: Celia had accused Willie of not having been clear about some money and he went at her like a man possessed. They exchanged noisy insults and I wasn't able to soothe their rage or get them to lower their voices and work out a solution on reasonable terms. In only a few minutes the tone had escalated to the level of a street brawl, which Nico finally stopped with the only yell any of us had heard in his lifetime and that paralyzed us with surprise. Willie exited with the slam of a door that nearly brought down the house. In one of the rooms, Tabra, still dazed from the effects of the operation and the painkillers, heard the screams and thought she was dreaming. Abuela Hilda and Sally, who was visiting, disappeared with the children—I think they hid in the cellar among the plaster skulls and the hidey-holes of the skunks.

Celia was acting in what she believed was my best interest, and I failed to jump in and go to my husband's defense, so the suspicion she had unleashed was left floating without being resolved. I never imagined that their argument was going to have such far-reaching consequences. Willie was deeply wounded, not by Celia, but by me. When finally we could talk about it, he said that I excluded him, that I formed an impenetrable circle with my family and left him outside. I didn't even trust him. I tried to smooth things over, but it was impossible. We had sunk pretty low. Willie and I didn't speak for several days, and we harbored a grudge for weeks. This time I couldn't run away because Tabra was convalescing with us, and my entire family was there in the house. Willie built a wall around himself; he was silent, furious, preoccupied. He went to the office early and returned late; he sat by himself to watch television, and stopped cooking for us. We ate rice and fried eggs every day. Not even the children could get through to him; they went around on tiptoes and got tired of inventing reasons to go to him: Grandpa had turned into an old grump. Nonetheless, we held to our agreement not to speak the word *divorce*, and I think that despite appearances, we both knew that we hadn't come to the end of our rope, we still had a lot in reserve. At night we slept on our own corner of the bed, but we always woke up with our arms around each other. Over time that helped us toward a reconciliation.

I may have given the impression in this account that Willie and I did nothing but argue, Paula. Of course it wasn't like that. Except for the times I went off to sleep at Tabra's, that is, at the most heated moments of our skirmishes, we went hand in hand. In the car, on the street, everywhere, always holding hands. It was like that from the first, but within two weeks of our meeting that custom became a necessity because of the affair of the shoes. Given my height, I have always worn high heels, but Willie insisted that I should be comfortable and not like the old-time Chinese women with their painful bound feet. He gave me a pair of athletic shoes that still today, eighteen years later, sit like new in their box. To please him, I bought a pair of sandals I saw on television. They had shown some slender models playing basketball in cocktail dresses and high heels . . . just what I needed. I threw away the shoes I'd brought from Venezuela and replaced them with those prodigious sandals. They didn't work, I just kept walking out of them. I was so often flat on my face on the floor that for reasons of basic safety

Willie has always grasped my hand tight. Besides, we're fond of each other and that helps whatever the relationship. I like Willie a lot, and I show that in a number of ways. He has begged me not to translate into English the love words I say to him in Spanish because they sound suspicious. I always remind him that no one has ever loved him more than I do, not even his mother, and that if I die he will end his days alone in some home for old folks, so it's worth his while to spoil me and celebrate me. Willie is not a man to squander romantic words, but if he has lived with me so many years without strangling me, it must be that he likes me a little. What is the secret of a good marriage? I can't say, every couple is different. The two of us are bound together by our ideas, a similar way of looking at the world, camaraderie, loyalty, humor. We look after each other. We have the same schedule, we sometimes use the same toothbrush, and we like the same movies. Willie says that when we're together our energies are multiplied, that we have the "spiritual connection" he felt when he first met me. It may be. All I know is that I like sleeping with him.

In view of our difficulties, we decided to have individual therapy, and Willie found a psychiatrist he got along with from the beginning, a huge bearded bear of a man that I perceived as my declared enemy, but one who with time would play a fundamental role in our lives. I don't know what Willie wanted to resolve in his therapy; I suppose the most urgent thing was his relationship with his children. In mine, when I began to rake through my memory, I realized that I was carrying a very heavy burden. I had to confront old silences, admit that my father's having abandoned me at the age of three had marked me and that the scar was still visible. It had determined my feminist posture and my relations with men from my grandfather and Tío Ramón, whom I had always rebelled against, to Nico, whom I treated as if he were a little boy—to say nothing of lovers and husbands to whom I had never completely given myself. In one session, my Buddhist therapist tried to hypnotize me. He didn't succeed, but at least I relaxed and I could see inside my heart an enormous black granite stone. I knew then that my task would be to rid myself of it. I would have to chip away at it, piece by piece.

To free myself of that dark rock, in addition to therapy and walks in the misty forest of your ashes, I took yoga classes, and I increased the number of calming acupuncture sessions with Dr. Shima, as much for the benefit of his presence as for his science. Lying on his cot with needles all over my body, I slipped away to other dimensions.

I was looking for you, daughter. I thought about your soul, trapped in an inert body through all that long year of 1992. At times I felt a claw in my throat, I could barely breathe, or I was weighed down by the weight of a sack of sand in my chest and felt as if I were buried in a deep hole, but soon I would remember to direct my breathing to the site of the sorrow, with calm, as it is thought one should do while giving birth, and immediately the anguish would diminish. Then I would visualize a stairway that allowed me to climb out of the hole and reach daylight, the open sky. Fear is inevitable, I have to accept that, but I cannot allow it to paralyze me. Once I said—or wrote somewhere—that after your death I was no longer afraid of anything, but that isn't true, Paula. I am afraid to lose persons I love or to see them suffer; I fear the deterioration of old age, I fear the world's increasing poverty, violence, and the world's corruption. In these years without you I have learned to manage sadness, making it my ally. Little by little your absence and other losses in my life are turning into a sweet nostalgia. That is what I am attempting in my stumbling spiritual practice: to rid myself of the negative feelings that prevent walking with assurance. I want to transform rage into creative energy and guilt into a mocking acceptance of my faults; I want to sweep away arrogance and vanity. I have no illusions, I

will never achieve absolute detachment, authentic compassion, or the state of ecstasy known to the enlight-
ened; it seems I do not have the bones of a saint, but I
can aspire to crumbs: fewer bonds, a bit of affection for
others, the joy of a clean conscience.

It's a shame you couldn't appreciate Miki Shima
during those months of his frequent visits to give you
Chinese herbs and acupuncture treatments. You would
have fallen in love with him, just the way my mother and
I did. He wears the suits of a duke, starched shirts, gold
cuff links, silk neckties. When I met him he had black
hair, but now, just a few years later, he is beginning
to show threads of gray, though he still doesn't have a
single wrinkle and his skin is as pink as an infant's, all
thanks to his miraculous ointments. He told me that his
parents lived together for sixty years, openly detesting
each other. The husband never spoke in the house and
the wife, specifically to exasperate him, never stopped,
but she served him like a good Japanese wife of that
time: she prepared his bath, scrubbed his back, put
food into his mouth, fanned him on summer days, "So
he could never say that she had failed in her duties," and
in the same manner he paid the bills and slept at home
every night, "So she could never say he was heartless."
One day the woman died, even though he was much
older than she and by rights he should have had lung

cancer because he smoked like a locomotive. She, who was strong and untiring in her hatred, was dispatched in two minutes by a heart attack. Miki's father had never so much as boiled water for his tea, much less washed his socks or rolled up the mat he slept on. His children expected him to waste away and die, but Miki prescribed some herbs for him and soon he began to put on weight and stand up straight, and laugh and talk for the first time in years. Now he rises at dawn, eats a ball of rice with tofu and the famous herbs, meditates, chants, does his tai chi exercises, and goes off with three packs of cigarettes in his pocket to catch trout. The walk to the river takes a couple of hours. He returns with a fish that he himself cooks, seasoned with Miki's magical herbs, and ends his day with a hot bath and a ceremony to honor his ancestors and, in passing, affront his wife's memory. "He is eighty-nine years old and he's fit as a pup," Miki told me. I decided that if those Chinese herbs had made that Japanese grandfather young again, they could dissolve the frightful rock in my heart.

Ballroom Dancing and Chocolate

One of the psychologists—we had several at our disposal—recommended that Willie and I share some activities that were fun, not just obligatory. We needed more lightness and entertainment in our lives. I proposed to my husband that we should take dancing lessons because we'd seen an Australian movie on that theme, *Strictly Ballroom*, and I could already see us whirling in the glow of crystal chandeliers, Willie in a dinner jacket and two-tone shoes and I in my beaded dress and ostrich plumes, both of us airy, graceful, moving to the same rhythm, in perfect harmony, as we hoped some day our relationship would be. When we had met that unforgettable day in October 1987, Willie had taken me dancing at a hotel in San Francisco. That gave me the opportunity to bury my nose in his chest and sniff him, and that was why I fell in love with him. Willie smells like a healthy boy. His only memory of that occasion, however, was that I kept tugging him around. It was like trying to break a wild mare, he told me. It seems to me that he asked, "Is this going to be a problem between us?" And he assures me that I answered, in a submissive little voice, "Of course not." That had been a number of years before.

We decided to begin with private lessons, so we wouldn't look ridiculous in front of other more advanced dancers. More accurately, I was the one who made that decision. The truth is that Willie was a good dancer in his youth, fawned over and with a winning record in dance contests, and I, in contrast, had all the grace of a Mack truck on the dance floor. The ballroom of the academy had floor-to-ceiling mirrors on all four sides, and the teacher turned out to be a nineteen-year-old Scandinavian with legs as long as I was tall. She was wearing black stockings with seams down the back, and stiletto-heeled sandals. She announced that we would begin by dancing the salsa. She pointed me to a chair, fell into Willie's arms, and waited for the precise beat of the music to launch herself across the floor.

"The man leads," was her first lesson.

"Why?" I asked.

"I don't know, but that's how it is," she said.

"Aha!" Willie crowed with a triumphant air.

"That doesn't seem fair to me," I persisted.

"What's not fair about it?" asked the Scandinavian.

"I think we should take turns. Willie will lead once and then it will be my turn."

"The man *always* leads!" the brutish woman exclaimed.

She and my husband glided around the dance floor to a Latin beat as the huge mirrors reflected their interlocked bodies to infinity, the long black-stocking-clad legs and Willie's idiotic smile, while I sat and stewed in my chair.

After we left the class, we had a fight in the car that came close to ending in fisticuffs. According to Willie, he hadn't even noticed the teacher's legs *or* her breasts. That was all in my head. "Jeez! How stupid can you get!" he cried. The fact that I had spent an hour in that chair while he danced was logical, since the man leads, and once he learned, he could guide me around the floor with the perfection of the courting dance of the heron. He didn't put it exactly that way, but to me it sounded as if he was mocking me. My psychologist thought that we should not give up, that ballroom dancing was an effective discipline for body and soul. What did he know, a Buddhist green tea drinker who surely had never danced in his life! But all the same, we went to a second and then a third class before I lost patience and punched the instructor. I have never felt more humiliated. The result was that what little we knew about dancing we lost, and since then Willie and I have gone dancing only once. I am recounting this episode only because it is like an allegory of our character: it captures every nuance from head to toe.

Celia, Nico, and the children moved into their new house, and Celia's brother came to live with them. He was a tall, pleasant, although rather spoiled young man who was looking for his destiny and planned to live in the United States. I believe that like Celia he had never gotten along very well with his family.

In the meantime, the publication of *Paula* brought me undeserved prizes and honorary degrees. I was elected a member of assorted academies of the language, and even given the symbolic keys to a city. Caps and gowns piled up in a trunk, and Andrea used them for her costumes—my granddaughter had entered a conservationist stage, and had a doll she named Save-the-Tuna. Luckily I never lost sight of something Carmen Balcells told me. "The prize doesn't honor the one who receives it as much as the one who gives it, so don't get any big ideas about yourself." That was impossible. My grandchildren made sure that I remained humble, and Willie reminded me that resting on your laurels is the best way to crush them.

It was about that time that Willie, Tabra, and I went to Chile to attend the premiere of *The House of the Spirits*. There were still Pinochet sympathizers in that country who were not ashamed to admit it. Today there are fewer in number because the general lost prestige among his faithful when the story of his thievery, his tax evasion, and his corruption came to light. The same people who had overlooked torture and murders could not forgive the millions he had stolen. It had been almost six years since the dictator had been defeated in a plebiscite, but the military, the press, and the judicial system still treated him with kid gloves. The right controlled the Congress, and the country was run under a constitution Pinochet himself had created; he counted on immunity from the Senate and the shelter of an amnesty law. The democracy was conditional, and there was a tacit social and political agreement not to provoke the military. A few years later, when Pinochet was arrested in England, where he had gone to collect his commissions on arms sales, have a medical checkup, and take high tea with his friend Margaret Thatcher, he was exposed in the world press and accused of crimes against humanity, and the legal edifice he had constructed to protect himself came tumbling down, and at last Chileans dared come out in the street and make fun of him.

The movie was as welcome as the plague among the extreme right, but it was enthusiastically received by most people, particularly by the young who had been raised under strict censorship and who wanted to know more about what had happened in Chile during the '70s and '80s. I remember that during the performance, one senator of the right jumped up and stamped out of the theater, announcing at the top of his lungs that the film was a string of lies besmirching a great patriot, our General Pinochet. Some reporters asked me what I thought about that, and I answered in good faith, since I had heard it said many times: "Everyone knows that gentleman is soft in the head." I regret that I've forgotten the man's name. . . . In spite of the few snags at the beginning, the film was very successful and now, ten years later, it is still one of the favorites on television and video.

Tabra, who though she had been to the least-known corners of the planet had never been in Chile, took away a very good impression of my country. I don't know what she had expected but she found herself in a city that reminded her of Europe, surrounded by magnificent mountains, hospitable people, and delicious food. We had a suite in the most luxurious hotel in town, where each night we were left a chocolate sculpture modeled on some aspect of our indigenous past, such as the *cacique* Caupolicán armed with a lance and followed by two or three of his Mapuche warriors. Tabra worked hard to eat the last crumb, with the hope of getting rid of it once and for all, but within a few hours it would be replaced with another two pounds of chocolate: a cart with two oxen, or six of our cowboys on horseback, the celebrated *huasos*, carrying the Chilean flag. And since she had learned as a child never to leave anything on her plate, she would give a great sigh and attack the plate, until the night she was conquered by a replica of Aconcague, the highest peak in the cordillera of the Andes in solid chocolate, as massive as the huge dark rock that according to my psychiatrist was sitting in the middle of my chest.

Children, Those Pint-Sized Lunatics

Willie and I realized with surprise that we had been together for nine years and that we were by now on a much firmer footing. According to him, he had from the first moment felt that I was his soul mate and had accepted me unconditionally, but that wasn't true in my case. Still today, a thousand years later, I marvel at the fact that we two met in all the world's vastness, felt attracted to each other, and managed to sweep away problems that at times seemed insurmountable, to form a couple.

The grandchildren, those pint-sized lunatics, as the Argentine humorist Quino defines them, have been the most entertaining part of our lives. The shadows around Sabrina's birth had cleared, and the gift her fairy godmother had given her as compensation for her physical limitations was now evident: a strength of character capable of overcoming obstacles that would have paralyzed a samurai. What other children did effortlessly, like walking or putting a spoonful of soup into their mouths, demanded invincible tenacity on her part, and she always achieved it. She limped, because her legs did not perform properly, but no one doubted that she was going to walk in the future, just as she had learned to swim, hang from a tree with one hand, and ride her bicycle pedaling with only one leg. Like her maternal grandmother, Willie's first wife, she was an extraordinary athlete. The upper part of her body was so strong and she was so agile that she was playing basketball in a wheelchair. She was a beautiful and delicate child the color of burnt sugar, with the profile of the famous queen Nefertiti. She learned to speak very early on and never showed the slightest trace of shyness, perhaps because she lived surrounded by people.

Alejandro was reminiscent of Nico in character, although he looked like his mother. Like his father, he had a curious mind and understood mathematical concepts before he could pronounce all the consonants in the alphabet. He was such a handsome boy that people stopped us in the street to rave about him. On one April 2, I remember the date very well, we were alone in the house and he came into the kitchen where I was making soup. He was frightened, and he threw his arms around my legs, and told me that there was "someone on the stairs." We went to look, went all through the house without finding anyone, and when we came back from the basement to the first floor he planted himself at the bottom of the stairs and pointed.

"Up there!"

"What is it, Alejandro?" I asked. All I saw were the tiled steps.

"She has long hair," and he hid his face in my skirts.

"It must be your aunt Paula. Don't be afraid, she's only come to say hello."

"She's dead!"

"It's her spirit, Alejandro."

"You told me she was in the woods! How did she get here?"

"In a taxi."

I supposed that by then you must have vanished, because he agreed to take my hand and go upstairs. I think the legend of the ghost began when my mother—who visited us a couple of times a year and stayed for several weeks since the trip between Santiago to San Francisco is a journey worthy of Marco Polo and can't be taken on lightly—told us she heard things at night, something like furniture being dragged. We all had heard it and we had different explanations: deer had got in and were walking around the terrace; it was the pipes contracting with the cold; the dry wood in the house was creaking. My friend Celia Correas Zapata, a professor of literature who had taught my books for years at San Jose University and was writing a book about my work entitled *Life and Spirit*, once stayed overnight. She slept in your room, Paula, and was awakened at midnight by the intense scent of jasmine, even though it was the dead of winter. She, too, mentioned noises, but no one gave her report much importance until a German journalist who stayed with us while he conducted a long interview swore that he had seen the bookcase slip noiselessly about a foot and a half from the wall, without disarranging the books. There was no earthquake that night, and this time it wasn't the perceptions of Latina women but the observation of a German male, whose opinion carried atomic weight. We accepted the idea that you came to visit us, although that possibility made the woman who cleaned the house very nervous. When Nico heard what had happened with Alejandro, he concluded that surely the boy had heard us talking and the rest was childish imagination. My son always has a rational explanation that ruins my best stories.

Andrea learned to tolerate her glasses and we were able to remove the rubber bands and safety pins, but her legendary clumsiness did not go away. She was always a little off center as she made her way through the world; she couldn't handle escalators or revolving doors. At the end of a school pageant, in which she appeared dressed as a Hawaiian girl with a ukulele, she made a long deep bow, but with her bottom turned to the audience. A unanimous bellow of laughter welcomed that irreverent salute, to the fury of the family and horror of my granddaughter, who was so embarrassed she wouldn't leave the house for a week. Andrea had the strange little face of a stuffed animal, accentuated by her curly hair. She was always in costume. She wore one of my nightgowns for an entire year, pink of course, and we have a photograph of her in kindergarten in a fur stole, a gift wrap ribbon on her chest, a bride's long gloves, and two ostrich plumes in her hair. She talked to herself because she heard the people in her stories, who wouldn't leave her in peace, and she was sometime frightened by her own imagination. There was a wall mirror in our house at the end of a hallway, and she often asked me to go with her "down the mirror road." As we neared the looking glass, her steps grew more hesitant because a dragon was crouching there, but just when the beast was ready to spring, Andrea would return from the other dimension to reality. "It's just a mirror, there's no monster there," she would tell me, though without much conviction. An

instant later she would be back in her story again, holding my hand along the imaginary road. "This child will
end up either stark raving mad or writing novels," her
mother decided. I was like her at that age.

Nicole shot up as soon as she began to walk, and from
being solid and square as an Inuit she went to floating
about with ethereal grace. Her mind was very sharp,
and she had a good memory, a sense of direction that
allowed her always to know where she was, and she could
have entranced Dracula with her round eyes and bunny
rabbit smile. Willie did not escape his granddaughter's
beguiling ways. Nicole had a mania for sitting beside
him while he was watching the news on television, but
after thirty seconds she would have convinced him that
cartoons would be better. Willie would go to another
room to watch his program, and Nicole, who hated to
be alone, would follow right behind. This routine would
be repeated several times during the evening. Once she
saw a male elephant mounting a female.

"What are they doing, Willie?"

"Coupling, Nicole."

"What?"

"They're making a baby."

"No, Willie, you don't understand, they're fighting."

"Okay, Nicole, they're fighting. Can I watch the
news now?"

At that moment a shot of a newborn elephant flashed on the screen. Nicole yipped, ran closer to look, stuck her nose against the screen, then turned toward Willie with her fists on her hips.

"That's what happens when you go around fighting, Willie!"

Nicole had to go to day care when she was still in diapers because all the adults in the family were working and couldn't look after her. Just the opposite of her sister, who always dragged along a suitcase containing her most precious treasures—an infinity of trinkets, the list of which was engraved in her memory—Nicole had absolutely no sense of possession; she was as free and unbound as a cricket.

Plumed Lizard

Tabra, the adventurer of our tribe, traveled several times a year to remote places, especially ones the State Department considered out of bounds to Americans, whether because of danger, as in the case of the Congo, or for being at political extremes, like Cuba. She had explored the world in various directions, almost always in primitive conditions, with the modesty of a pilgrim, and alone . . . until she met the man who was willing to accompany her. As I have lost count of my friend's suitors, and some anecdotes are hazy in my memory, all of Tabra's lovers from that time have blended in my mind into a single character. Let's call him Alfredo López Lagarto-Emplumado, or Plumed Lizard. Picture a very smart man, so handsome that he never lost an opportunity to look at himself in every mirror and plate glass window he passed. With olive skin and a slim, muscular body, he was a treat to the eyes, especially to Tabra's, who sat mute with admiration while he talked about himself. His father was a Mexican from Cholula and his mother a Comanche Indian from Texas, which guaranteed him a lifetime of thick black hair, which he usually wore in a ponytail, unless Tabra had braided it and adorned it with beads and feathers. He had always been curious about traveling but hadn't been able to do it because of his limited income.

Lagarto-Emplumado had been preparing all his life for a very noble and secret mission that, though secret, he told to anyone who would listen: it was to rescue the crown of Moctezuma displayed in a museum in Austria and return it to the Aztecs, its legitimate owners. He had a black shirt bearing the legend "Crown or Death, Viva Moctezuma!" Willie wanted to know if the Aztecs had given their approval of his project and he told us they hadn't; his plan was still very secret. The crown, made of four hundred quetzal plumes, was more than five centuries old, and it was possible that it might be a bit moth-eaten. At a family dinner, we asked how he meant to transport it and he never came to visit again; maybe he thought we were making fun. Tabra explained to us that imperialists appropriate the cultural treasures of other nations, the way the British did when they robbed Egyptian tombs. As for him, he admired the tattoo of Quetzalcoatl she had on her left ankle. It was not by chance, he said, that Tabra had a tattoo of the Mesoamerican god, the plumed serpent that had been the inspiration for his own name.

At the request of Lagarto, who like a good Comanche felt the call of the desert, they made a trip to Death Valley. I warned Tabra that it wasn't a good idea; even the name of the place was a bad omen. She drove for days, then, carrying the tent and supplies, plodded behind her hero for several miles, dehydrated and faint from heat exhaustion, as he picked up little sacred stones for his rituals. My friend kept her complaints to herself so he wouldn't throw her below-par physical condition-ing and her age in her face: she was twelve years older than he. Finally Lagarto-Emplumado found the perfect place to camp. Tabra, red as a beet, her tongue swollen, set up the tent and fell onto her sleeping bag, shivering with fever. The champion of the indigenous cause shook her to get up and fix him *huevos rancheros*. "Water, water . . . ," Tabra muttered. "Even if my mother was dying, she would have cooked my father's beans," her peevish Lagarto replied.

Despite the experience in Death Valley, where she came close to leaving her sun-baked bones, Tabra invited her Plumed Lizard to go with her to Sumatra and New Guinea, where she hoped to find inspiration for her ethnic jewelry and a shrunken head to add to her collection of rare objects. Lagarto-Emplumado, who was extremely careful of his physical condition, took along a tote heavy with lotions and ointments, which he shared with no one, and a handbook on all the illnesses and accidents that can befall a traveler on this planet, from beriberi to a python attack. In a village in New Guinea, Tabra fell prey to a cough; she was pale and exhausted, maybe a consequence of the brutal operation on her breasts.

"Don't touch me! You may be contagious. You may have an illness that comes from eating the brains of your ancestors," said Lagarto-Emplumado, highly alarmed, after consulting his encyclopedia of misfortunes.

"What ancestors?"

"Any ancestors. They don't have to be ours. These people eat the brains of the dead."

"They don't eat the whole brain, Lagarto, only a little bit, as a sign of respect. But I doubt that we've eaten any at all."

"Sometimes you can't tell what you have on your plate. Besides, we've eaten pork, and all the pigs in Bukatingi feed on what they can scavenge. Haven't you seen them rooting in the cemetery?"

Tabra's relationship with Alfredo López Lagarto-Emplumado was temporarily interrupted when he decided to return to a former lover, who had convinced him that only someone who was pure of heart could rescue Moctezuma's crown, and that as long as he was with Tabra, his was contaminated. "Why is she purer than you?" I asked my friend, who had contributed to the funds needed to carry out the epic of the crown. "Don't worry, he'll be back," Willie consoled her. God forbid! I thought, ready to tear the memory of that ingrate to shreds, but when I saw Tabra's misty eyes I thought it better to hold my tongue. Lagarto returned as soon as he realized that the other woman, no matter how pure, was not planning to finance him. He came back with the idea that they could form a love triangle, but Tabra would never have accepted such a Mormon-inspired solution.

It was about that time that my friend's former husband died, the preacher from Samoa. He weighed 330 pounds and had high blood pressure and galloping diabetes. They'd had to remove a foot and then several months later, amputate the leg above the knee.

Tabra has told me what she suffered in her marriage; I know that it took years of therapy to overcome the trauma produced by that man's violence, who had seduced her when she was a girl, convinced her that they should run away together, beaten her brutally from the first day, kept her terrorized for years, and then after the divorce turned his back on their son. Tabra raised Tangi alone, with no help of any kind from the boy's father. However, when I asked her if she was happy he'd died, she looked at me with surprise. "Why would I be happy? Tangi is sad, and he left many other children behind."

Life Comrade

Compared with Plumed Lizard, my life comrade, Willie, is a freaking wonder; he takes care of me. And compared to Tabra's expeditions to the farthest confines of the earth, my little professional trips were pitiful; even so, they left me drained. I had to board a plane day after day, where I struggled valiantly against the viruses and bacteria of the other passengers; I spent weeks away from home, and days preparing talks. I don't know how I stole time to write. I learned to speak in public without panicking, to go through airports without getting lost, to survive on what a carry-on would hold, to whistle down a taxi, and to smile at people greeting me, even though my stomach hurt and my shoes were too tight. I don't remember all the places I went, it doesn't matter. I know I traveled through

Europe, Australia, New Zealand, Latin America, parts of Africa and Asia, and all the states of the union except North Dakota. On planes, I wrote my mother by hand to tell her my adventures, but when I read the letters a decade later, it's as if all that had happened to another person.

The one vivid memory that stays in my mind was of a scene in New York in midwinter that would haunt me until later, following a trip to India, I was able to exorcise it. Willie had joined me for the weekend and we had just visited Jason and a group of his university friends, young intellectuals in leather jackets. In the months he'd been away from Sally, he'd not spoken of marriage again, and we had the idea that their engagement was ended. Sally herself had hinted at that on a couple of occasions, but Jason denied it. According to him, they were going to be married as soon as he graduated, but when Ernesto had visited us in California we'd learned that he'd had a brief but intense affair with Sally, so we assumed that she and Jason no longer had ties. Jason, incidentally, didn't learn about Sally and Ernesto until many years later. By then the events that demolished his faith in our family, which he had idealized, had already been set in motion.

Willie and I had said good-bye to his son with great emotion, thinking how much he had changed. When

I had come to live with Willie, Jason spent his nights reading or out partying with his buddies; he got up at four in the afternoon, threw a grimy coverlet around him, and settled on the terrace to smoke, drink beer, and talk on the phone until I rapped him on the head enough times that he went to class. Now he was on the way to becoming a writer, something we'd always thought he would do because he was very talented. Willie and I were remembering that stage of the past as we walked down Fifth Avenue in the midst of the noise and crowds and traffic and cement and frost. In front of a shop window exhibiting a collection of the ancient jewels of imperial Russia, we saw a woman huddled on the ground, shivering. She was of the African race, filthy, wearing rags topped with a black garbage bag. She was sobbing. People were hurrying by without looking at her. Her weeping was so desperate that for me the world froze, as in a photograph; even the air absorbed the fathomless pain of that wretched woman. I crouched down at her side and gave her all my cash, though I was sure that a pimp would soon be by to take it from her. I tried to communicate, but she didn't speak English—or else she was beyond words. Who was she? How had she arrived at such a state of desolation? Perhaps she'd come from a Caribbean island, or from the coast of Africa, and waves had haphazardly washed her onto Fifth Avenue the way that meteorites fall to earth from another dimension. I always wonder what could have become of her. I've never forgotten her, and I carry the terrible guilt that I couldn't or wouldn't help her. We kept walking, hurrying in the cold, and a few blocks later we were inside the theater and the woman was left behind, lost in the night. I never imagined then that I could never forget her, that her tears would be an inescapable call, until a couple of years later life would give me the opportunity to respond.

If Willie could manage to get away from work, he would fly and meet me at different points of the country so we could spend one or two nights together. His office kept him tied down and gave him more disappointments than satisfaction. His clients were down-and-out folks who'd been injured on the job. As the number of immigrants from Mexico and Central America, most of them illegal, increased in California, so did the xenophobia. Willie charged a percentage of the compensation he negotiated for his clients, or obtained in a trial, but those sums were getting smaller and smaller, and the cases difficult to win. Fortunately, he didn't pay rent since his office was housed in our erstwhile brothel in Sausalito. Tong, his accountant, performed juggling acts to cover salaries, bills, taxes, insurance, and banks. This noble Chinese man looked after Willie as he would a foolish son, and he cut so many corners that his frugality had reached the level of legend. Celia swore to us that at night, after we left the office, he pulled the paper cups from the trash, washed them, and put them back in the kitchen. The truth is that without the vigilant eye and the abacus of his accountant, Willie would have gone under.

Tong was almost fifty, but he looked like a young student: slim, small, with a mop of bristly hair, he always dressed in jeans and sneakers. He hadn't spoken to his wife for twelve years, though they had lived under the same roof; they hadn't divorced because they would have had to divide their savings. They were also afraid of his mother, a tiny, ferocious old lady who had lived in California for thirty years but believed she was in the south of China. This lady did not speak a word of English; she did all her shopping with merchants in Chinatown, listened to a Cantonese radio station, and read the San Francisco newspaper in Mandarin. Tong and I had in common our affection for Willie; that was a bond despite the fact that neither of us could understand the other's accent. At the beginning, when I had just come to live with Willie, Tong felt an atavistic distrust of me, which he made obvious at the slightest opportunity.

"What does your accountant have against me?" I asked Willie one day.

"Nothing in particular. All the women in my life have been expensive, and since he pays my bills, he would like for me to live in strict celibacy," he informed me.

"Tell him that I have supported myself since I was seventeen."

I suppose that he did, because Tong began to look at me with something like respect. One Saturday he found me in the office scrubbing the bathrooms and vacuuming; at that point his respect was transformed into open admiration.

"You marry this one. She clean," he counseled Willie in his rather limited English. He was the first to congratulate us when we announced we were going to be married.

This long love affair with Willie has been a gift of the mature years of my life. When I divorced your father, Paula, I prepared myself to go on alone, because I thought it would be next to impossible to find a new life companion. I'm bossy, independent, tribal, and I have unusual work habits that cause me to spend half my available time alone, not speaking, in hiding. Few men can cope with all that. But I don't want to commit the sin of false modesty, I also have a few virtues. Do you remember any, daughter? Let's see, let me think. . . . Well, for example, I'm low-maintenance, and I'm healthy and affectionate. You always said that

I'm entertaining and that no one would ever get bored
with me, but that was then. After I lost you, I also lost
my desire to be the life of the party. I've become introverted; you wouldn't recognize me. The miracle was
finding—where and when I least expected—the one
man who could put up with me. Synchronicity. Luck.
Destiny, my grandmother would have said. Willie main-
tains that we have loved each other in previous lives and
will continue to do so in future ones, but you know how
the idea of karma and reincarnation frightens me. I'd
rather limit this amorous experiment to a single life, for
that's enough. Willie still seems such a stranger to me!
In the morning, when he's shaving and I see him in the
mirror, I often ask myself who the devil that large, too
white, North American man is, and what are we doing
in the same bathroom. When we met we had very little
in common; we came from very different backgrounds
and we had to invent a language—Spanglish—in order
to understand each other. Past, culture, and customs
separated us, as well as the inevitable problems of
children in a family artificially glued together, but by
elbowing our way forward, we succeeded in opening
the space that is indispensable for love. It's true that
to make my life in the United States with Willie, I left
behind nearly everything I had, and adjusted however
I could to the disarray of his existence—but he had to

make his own concessions and changes in order for us to be together. From the beginning, he adopted my family and respected my work; he has accompanied me in every way he could; he has backed me up and protected me even from myself; he never criticizes me; he gently laughs at my manias; he doesn't let me run over him; he doesn't compete with me, and even in the fights we've had he acts with honor. Willie defends his territory, but without aggression; he says he had traced a small chalk circle around him, and within it he is safe from me and my tribe: be careful not to invade it. A great pool of sweetness lies just beneath the surface of his tough appearance; he is as sentimental as a big dog. Without him, I wouldn't be able to write as much and as calmly as I do because he takes care of all the things that frighten me, from my contracts and our social life to the functioning of all our mysterious household machines. Even though I am still surprised to find him by my side, I have become so used to his massive presence that now I couldn't live without him. Willie fills the house, fills my life.

The Empty Well

In the summer of 1996, an unhinged racist in Oklahoma City used a truck loaded with a thousand kilos of explosives to blow up a federal building. Five hundred people were wounded and one hundred and sixty-eight were killed, several children among them. One woman was trapped under a massive block of cement and they had to amputate her leg without anesthesia to save her. Celia sobbed over that for three days; she said that it would have been better had the poor woman died, since she not only lost her leg in the tragedy, she also lost her mother and her two small children. Celia's reaction was similar to those she'd had from other tragedies she'd read about, she had few defenses against the outside world; despite our long friendship I couldn't detect what was bothering her. I thought I knew Celia better than she knew herself, but there was a part of my daughter-in-law's soul that escaped me, as I realized a few weeks later.

Willie and I decided that it was time to take a vacation. We were exhausted and I could not shake off my grief, although it had been nearly four years since you died and three since Jennifer had disappeared. I didn't know then that the sadness is never entirely gone; it lives on forever just below the skin. Without it I wouldn't be who I am, or be able to recognize myself in the mirror. Ever since I finished *Paula,* I hadn't written a word of fiction. For years I'd been playing with the idea of a novel about the mid-nineteenth-century gold rush in California, but I wasn't enthusiastic enough to tackle such a long and demanding project. I was as active as always and few people suspected my state of mind, but deep in my soul I was moaning. I developed a taste for solitude; I wanted only to be with my family; people bothered me, my friends were reduced to three or four. I was spent. I didn't want to keep making tours to promote my books, explaining what I'd already said in those pages. I needed silence, but it became more and more difficult to find. Journalists came from all over and invaded us with their cameras and lights. On one occasion, some Japanese tourists came to observe our house as if it were a monument, just as a crew from

Europe had arrived hoping to photograph me inside an enormous cage with a majestic white cockatoo. That very large bird did not look at all friendly, and it had claws like a condor. It came with a trainer, and he should have controlled it, but it shit on all the furniture and when I went into the cage nearly poked out my eye. However, overall I really couldn't complain. I had an affectionate public and my books were being read everywhere. My sadness manifested itself in sleepless nights, dark clothing, the wish to live in a hermit's cave, and an absence of inspiration. I summoned the muses in vain. Even the most bedraggled muse had abandoned me.

For someone who lives to write and lives from what she does write, an internal drought is terrifying. One day I was in Book Passage, killing time with several cups of tea, when Anne Lamott came in; she is a North American writer much beloved for stories filled with humor and spirituality. I told her that I was blocked and she told me that the business of the "writer's block" is nonsense, and what happens is that sometimes the well has gone dry and has to be refilled.

The idea that my well of stories and my wish to tell them was drying up threw me into a panic, because no one was going to give me a job doing anything else, and I had to help support my family.

Nico had a job as a computer technician in a nearby city and was commuting more than two hours a day and Celia was doing the work of three people, but they couldn't meet costs for their children, we lived in one of the most expensive areas of the United States. Then I remembered that I was trained as a journalist; if I'm given a subject and time to research it, I can write about almost anything—except politics and sports. I assigned myself a "feature" as different as possible from my last book, one that had nothing to do with pain and loss, the pleasureful sins of life: gluttony and lust. As it would not be fiction, the caprices of the muse had little bearing; all I had to do was my research on food, eroticism, and the bridge that connected them: aphrodisiacs. Calmed by that plan, I accepted Willie and Tabra's suggestion that we go to India, although I had no desire to travel, and even less to India, the farthest possible point from our home before starting back around the other side of the planet. I didn't think I would be able to bear the legendary poverty of India, the devastated villages, starving children, and nine-year-old girls sold into early marriages, forced labor, or prostitution, but Willie and Tabra promised me that India was much more than that, and they were determined to take me if they had to tie me up to do it. Besides, Paula, I had promised you that one day I would visit that country because you had come back from a trip there fascinated, and you convinced me that India is the richest source of inspiration for a writer. Alfredo López Lagarto-Emplumado did not come with us, though he was again visible on Tabra's horizon; he was planning to spend a month communing with nature, accompanied by a pair of Comanches, tribal brothers. Tabra had to buy him some sacred drums that apparently were indispensable for their rituals.

Willie bought a khaki explorer's outfit with thirty-seven pockets, a backpack, an Aussie hat, and a new lens for his cameras, about the size and weight of a small cannon. Tabra and I packed our usual Gypsy skirts, ideal because wrinkles and stains wouldn't show. The three of us set off on a journey that ended a century later when we landed in New Delhi and sank into the city's sticky heat and its cacophony of voices, traffic, and blasting radios. We were surrounded by a million hands, but fortunately Willie's head emerged like a periscope above the mass of humanity, and in the distance saw a sign with his name held by a tall man with a turban and authoritative mustache. It was Sirinder, the guide we had hired through an agency in San Francisco. He opened a way with his cane, chose some bearers to carry the luggage, and took us to his ancient automobile.

We stayed in New Delhi several days. Willie was agonizing with an intestinal infection and Tabra and I were roaming around buying bagatelles. "I think your husband is pretty sick," she told me the second day, but I wanted to go to the quarter where the craftsmen who carved stones for her jewelry had their shops. The third day Tabra pointed out to me that Willie was so weak that he wasn't even talking, but as we hadn't as yet visited the street of the tailors, where I wanted to buy a sari, I didn't take immediate action. I conjectured that what Willie needed was time; there are two kinds of illness: the ones that simply go away and the deadly ones. That night Tabra suggested that if Willie died, it might ruin our trip. Faced with the possibility of having to cremate him on the banks of the Ganges, I called the hotel desk and they soon sent up a doctor: short, oily hair, wearing a shiny brick-colored suit. When he saw my husband looking like a corpse, he did not seem in the least alarmed. He pulled from his battered doctor's bag a glass syringe like the one my grandfather used in 1945, and prepared to inject the patient with a viscous liquid; the needle was resting in a cotton ball and to every appearance was as ancient as the syringe. Tabra wanted to intervene, but I assured her that there was no need to make a fuss over a possible case of hepatitis when the future of the patient was uncertain anyway.

The doctor worked the miracle of restoring Willie to good health in twenty hours, and so we were able to continue our journey.

India was one of those experiences that mark you for life, memorable for many reasons, though as this is not a travelogue it isn't the place to recount them. I will relate only two relevant episodes. The first gave me the idea of a way to honor your memory, daughter, and the second changed our family forever.

Who Wants a Girl?

Sirinder, our driver, had the expertise and the daring needed to move through the city traffic, dodging cars, buses, burros, bicycles, and more than one starving cow. No one hurried—life is long—except the motorcycles zigzagging at the speed of torpedoes and with five riding aboard. Sirinder showed signs of being a man of few words, and Tabra and I learned not to ask him questions because the only one he answered was Willie. The rural roads were narrow and curving, and he drove them at breakneck speed. When two vehicles met nose to nose, the men at the wheel looked each other in the eye and determined in a fraction of a second which was the alpha male, then the other man ceded right-of-way. The accidents we saw always involved two trucks of similar size that had smashed head-on because it wasn't clear in time which was the alpha driver. We didn't have safety belts, we had karma; no one dies before his time. We did not drive with lights at night for the same reason. Sirinder's intuition warned him that a vehicle might be coming toward us, at which time he flashed on his lights and blinded the driver.

As we drove out from the city, the landscape became sere and golden, then dusty and reddish. The villages were farther and farther apart, and the plains stretched forever, but there was always something to attract our attention. Willie carried his camera bag, tripod, and cannon-sized lens everywhere, a rather complex apparatus to set up. It is said that the only thing a good photographer remembers is the photo he didn't take. Willie will remember a thousand, like an elephant painted with yellow stripes and dressed as a trapeze artist, all by itself in that open countryside. On the other hand, he was able to immortalize a group of workers who were moving a mountain from one side of the road to the other. The men, wearing nothing but loin cloths, were piling rocks into the baskets the women carried across the road on their heads. The women were graceful, slim, dressed in threadbare saris of brilliant colors— magenta, lime, emerald—and they moved like reeds in the wind, carrying their burden of rocks. They were classified as "helpers," and they earned half of what the men did. When it was time to eat, the men squatted in a circle, holding their tin plates, and the women waited a respectful distance away. Later they ate anything the men left.

After still more hours of driving we were tired; the sun was beginning to go down and brushstrokes the color of fire streaked the sky. In the distance, in the dry fields, stood a solitary tree, perhaps an acacia, and beneath its branches we could see some dark figures that looked like huge birds but as we went closer turned out to be a group of women and children. What were they doing there? There wasn't any village or well nearby. Willie asked Sirinder to stop so we could stretch our legs. Tabra and I walked toward the women, who started to back away, but their curiosity overcame their shyness and soon we were together beneath the acacia, surrounded by naked children. The women were wearing dusty, frayed saris. They were young, with long black hair, dry skin, and sunken eyes made up with kohl. In India, as in many parts of the world, the concept of personal space we defend so fiercely in the West doesn't exist. Lacking a common language, they greeted us with gestures, and then they examined us with bold fingers, touching our clothing, our faces, Tabra's red hair, something they may not have seen before, and our silver jewelry. We took off our bracelets and offered them to the women, who put them on with the delight of teenagers. There were enough for everyone, two or three each.

One of the women, who could have been about your age, Paula, took my face in her hands and kissed me lightly on the forehead. I felt her parted lips, her warm breath. It was such an unexpected gesture, so intimate, that I couldn't hold back the tears, the first I had shed in a long time. The other women patted me in silence, disoriented by my reaction.

From the road, a toot of the horn from Sirinder told us that it was time to leave. We bade the women good-bye and started back to the car, but they followed us. One touched my shoulder. I turned, and she held out a package. I thought she meant to give me something in exchange for the bracelets, and tried to explain with signs that it wasn't necessary, but she forced me to take it. It weighed very little, I thought it was a bundle of rags, but when I turned back the folds I saw that it held a newborn baby, tiny and dark. Its eyes were closed and it smelled like no other child I have ever held in my arms, a pungent odor of ashes, dust, and excrement. I kissed its face, murmured a blessing, and tried to return it to its mother, but instead of taking it, she turned and ran back to the others, while I stood there, rocking the baby in my arms, not understanding what was happening. A minute later Sirinder came running and shouting to put it down, I couldn't take it, it was dirty, and he snatched it from my arms and started toward the women to give it back, but they ran away, terrified by the man's wrath. And then he bent down and laid the infant on the dry earth beneath the tree.

By that time, Willie had come too, and he hustled me back to the car, nearly lifting me off the ground, followed by Tabra. Sirinder started the engine and we drove off, as I buried my head in my husband's chest.

"Why did that woman try to give us her baby?" Willie murmured.

"It was a girl. No one wants a girl," Sirinder explained.

There are stories that have the power to heal. What happened that evening beneath the acacia loosened the knot that had been choking me, cleaned away the cobwebs of self-pity, and forced me to come back to the world and transform the loss of my daughter into action. I could not save that baby girl, or her desperate mother, or the "helpers" who were moving a mountain rock by rock, or millions of women like them and like the unforgettable woman I saw crying on Fifth Avenue that winter in New York, but I promised at that moment that I would at least attempt to ease their lot in life, as you would have done. For you, no act of compassion was impossible. "You have to earn a lot of money with your books, Mamá, so I can start a shelter for the poor and you can pay the bills," you told me one day, entirely serious. The money I had made, and was still making, from the publication of *Paula* was sitting in a bank waiting for me to decide how to use it. At that moment, I knew. I calculated that if the capital would grow with every book I wrote in the future, something good would come of it: only a drop of water in the desert of human need, but at least I wouldn't feel helpless. "I am going to establish a foundation to help women and children," I told Willie and Tabra that night, never imagining that with the years that seed would become a tree, like the acacia.

A Voice in the Palace

The palace of the maharajah, all gleaming marble, stood in a Garden of Eden where time did not exist, the climate was always gentle, and the air carried the scent of gardenias. Water from the fountains ran along sinuous canals among flowers, golden birdcages, white silk parasols, and majestic peacocks. The palace was now owned by an international hotel chain that had had the good judgment to preserve the original charm. The maharajah, ruined, but with dignity intact, occupied a wing of the building, protected from the curiosity of outsiders by a screen of palms and purple bougainvillea. In the calm of the afternoon he liked to sit in the garden and have tea with a girl who had not yet reached puberty, and who was not his great-granddaughter but his fifth wife. That interlude was assured by two guards in imperial uniforms and plumed turbans, with scimitars at the waist. In our profusely decorated suite, worthy of a king, there was not one inch where you could rest your eyes. From our balcony we had a view of the entire garden, which was separated by a high wall from the neighborhoods of the poor stretching as far as the horizon. After traveling dusty roads for weeks, we could rest in this palace, with its army of silent employees to carry our clothing to be washed, bring us tea and honey cakes on silver trays, and prepare our foaming baths. It was paradise. We dined on the delicious cuisine of India, which Willie was already immunized against, and fell into bed disposed to sleep forever.

The telephone rang at three in the morning—the time indicated by the green numbers on the travel clock glowing in the darkness—waking me from a hot, heavy sleep. I put out my hand, feeling for the phone, finding nothing, until my fingers touched a switch and I turned on the lamp. I didn't know where I was, or what the transparent veils floating above my head were, or the winged demons threatening me from the painted ceiling. I was aware of moist sheets stuck to my skin and a sweet scent I couldn't identify. The telephone kept ringing, and with every jangle my apprehension grew; it had to be something calamitous to justify the urgency of calling at that hour. Someone died, I said aloud. Be calm, be calm, I told myself. It couldn't be Nico. I had already lost a daughter and according to the law of probability I would not lose another child in my lifetime. And it wasn't my mother, she's immortal. Maybe there was news about Jennifer. Had she been found? The continued ringing guided me to the far end of the room, where I discovered an antiquated telephone sitting between two porcelain elephants. From the other side of the world, with the clarity of an omen, came the unmistakable voice of Celia. I couldn't find the strength to ask her what had happened.

"It seems that I'm bisexual," she announced in a quavering voice.

"What is it?" Willie asked, dazed with sleep.

"Nothing. It's Celia. She says she's bisexual."

"Oh!" My husband snorted and fell back to sleep.

I suppose that Celia called to ask me for help, but I could think of nothing magical that would help at that moment. I begged my daughter-in-law not to rush and do anything desperate, since we are all more or less bisexual and if she had waited twenty-nine years to discover that, she could wait until we returned to California. A matter as important as this should be discussed within the family. I cursed the distance that prevented me from seeing the expression on her face. I promised that we would come back as quickly as possible, although at three in the morning there wasn't much we could do to change our airline tickets, a process that even by day was complicated in India. The call had killed any chance of sleeping, and I did not go back to the veil-draped bed. Neither did I dare wake Tabra, who was in a different room on the same floor.

I went out on the balcony and waited for morning in a polychrome wood swing with topaz-colored silk cushions. A climbing jasmine and a tree with large white flowers were releasing that courtesan's fragrance I had noted in our room. Celia's news had produced a rare lucidity. It was as if I could see my family from above, floating overhead. "This daughter-in-law of ours never fails to surprise me," I murmured. In Celia's case, the word *bisexual* could have several connotations, but none would be without pain to my people. Hmmm. Without thinking, I wrote *my. . . .* That's how I feel about all of them; they all belong to me: Willie, my son, my daughter-in-law, my grandchildren, my parents, and even my stepchildren, with whom I lived from skirmish to skirmish . . . they're all mine. It had been an effort to bring them together, and I was prepared to defend that small community against the vagaries of fate and bad luck. Celia was an uncontainable force of nature; no one had any influence over her. I didn't ask myself twice whom she had fallen for, the answer was obvious to me. "Help us, Paula, this is no joke," I begged you, but I don't know whether you heard me.

Nothing Deserving Thanks

The disaster—I can think of no other word to describe it—unfolded at the end of November, Thanksgiving Day. It's true, that seems ironic, but we don't get to choose the dates for such episodes. We returned to California as quickly as we could, but to find flights, change the tickets, and fly across half the planet took more than three days. The night that Celia waked me, I'd told Willie what was going on, but he was asleep; he hadn't really heard me and I had to tell him again the next morning. It made him laugh. "That Celia is a loose cannon," he said, not considering the consequences my daughter-in-law's announcement would have for the family. Tabra had to go on to Bali, so we said good-bye without much explanation. When we got to San Francisco, Celia was waiting for us at the airport; we didn't, however, discuss anything until the two of us were alone. This was not a confidence she wanted to share in front of Willie.

"I never dreamed this was going to happen to me, Isabel. You remember what I always thought of gays," she told me.

"I remember, Celia. How could I forget? Have you gone to bed with her?"

"With who?"

"With Sally, who else?"

"How do you know it's her?"

"Oh, Celia, no need of a crystal ball for that. Did you sleep with her?"

"That isn't important!" she exclaimed with burning eyes.

"To me it seems very important, but I may be mistaken. . . . The heat of passion passes, Celia, and it's not worth destroying a marriage for. You're confused by the novelty, that's all."

"I am married to a marvelous man, and I have three children I will never live without. You can imagine how long I thought about this before I told you. You don't make a decision like this lightly. I don't want to hurt Nico and the kids."

"It's strange that you make your confession to me, I'm your mother-in-law. You don't think that unconsciously . . . ?"

"Don't come at me with your fucking psychology!" she interrupted. "You and I tell each other everything." And that was true.

I endured a week of brutal anxiety, but nothing compared with the weeks it had taken Celia and Sally to decide their future. They had lived in the same house, worked together, shared children, secrets, interests, and fun, but they were very different in character, and maybe that explained the mutual attraction. Abuela Hilda had pointed out to me earlier that "those girls love each other very much." Nothing got by that quiet, discreet, nearly invisible grandmother. Had she been trying to warn me? Impossible to know; that diplomatic woman would never have made a malicious comment.

Confused about carrying that secret, I debated with myself as I was preparing the Thanksgiving turkey. I was following a new recipe, one my mother had sent me in the mail. You put a pile of herbs in the blender with olive oil and lemon, then with a syringe inject the green mixture beneath the skin of the bird and let it marinate for forty-eight hours.

Sally had stopped coming to work in my office, but we saw each other nearly every day when I looked in on my grandchildren; she spent a lot of time in that house. I tried not to stare at her and Celia when they were together, but if they accidentally brushed against each other, my heart gave a leap. Willie, tired from the long trip to India and hung over from his intestinal infection, stayed on the fringe of things with the hope that passions would evaporate.

Luckily, I was able to get an appointment with my psychologist, whom I hadn't seen for a long time. He had moved to southern California but had come to San Francisco to spend the holidays with his family. We met at a coffee shop since he no longer had an office, and as he was sipping his green tea and I my cappuccino, I brought him up to speed on the family soap opera. He asked me if by any chance I was deranged. How had it occurred to me to act as the go-between in a situation like this? It wasn't a secret it was up to me to keep.

"You are the mother figure, in this case an archetype: mother of Nico, stepmother of Jason, mother-in-law of Celia, grandmother of the children. And future mother-in-law of Sally—if this hadn't happened," he explained.

"That I doubt. I don't think Sally would have married Jason."

"That isn't the point, Isabel. You must confront them and demand that they confess the truth to Nico and Jason. Give them twenty-four hours. If they don't do it, you will have to do it yourself."

I followed his advice. The twenty-four hours would be up precisely during the long weekend of Thanksgiving, a sacred holiday for Americans.

To celebrate the festive day, the family was going to get together for the first time in months, including Ernesto, who called and said he had fallen in love with a girl he worked with. Her name was Giulia, and he was bringing her to California to meet the family. This was not a propitious moment! He would arrive first from New Jersey and Giulia would appear the following day, which gave us a little time to tune everyone in. It was good that Fu, Grace, and Sabrina were having their dinner at the Zen Center, so there would be three fewer witnesses. Willie and I were so befuddled we couldn't help anyone. I don't know how we survived that horrendous weekend without violence. Celia locked herself away with Nico and I can't imagine how she told him; there was no diplomatic way to do it or to avoid the emotional anguish of her news. It would be impossible—as she feared—not to wound him and the children. I think that at first Nico did not fully realize the dimensions of what had happened, and believed that with imagination and tolerance things could be sorted out. Weeks, perhaps months, would go by before he understood that his life had changed forever.

Jason and Sally were separated not only by distance but also by the fact that they had little in common. It was difficult to imagine Sally living a nocturnal, bohemian life amid intellectuals in the chaos of New York, or Jason in California, vegetating in the bosom of the family and bored to death. Many years later, when I spoke with both of them about these events, I realized that their two versions were contradictory. Jason assured me that he was in love with Sally and convinced that they would get married, and because of that he had lost his head when she called to tell him what had happened.

"I have something to tell you," she announced. He immediately thought that she had been unfaithful and a wave of anger swept over him, though his next thought was that it wasn't anything too serious, since she was prepared to confess it. When she managed to get out the words to explain that it had to do with a woman, Jason drew a breath of relief: he was not actually facing a rival; this was a foolishness women do out of curiosity, but then Sally added that she was in love with Celia. With that dual betrayal, Jason felt as if he had been clubbed. He was not only losing someone he thought was still his sweetheart, he was also losing a sister-in-law he loved like a sister. He felt betrayed by everyone, including Nico, who hadn't stopped it from happening. He appeared that accursed weekend looking thin and drawn; he had lost I don't know how many pounds. He marched into the house with his knapsack on his back, unshaven, teeth clenched, smelling of alcohol and pale with rage. He had to deal with the situation without any support from anyone; all of us were lost in our own passions.

Sally picked up Ernesto at the airport. He was coming from New Jersey, where he'd lived since 1992, when we brought you, in a coma, to California. She took him to a café to warn him of what was going on. He could not just be dropped into the middle of the melodrama or he would think we had all gone mad. How would he explain all this to Giulia? His girlfriend was a tall, chatty blond with sky-blue eyes, with that freshness of people who have faith in life. We Sisters of Disorder had prayed for several years that Ernesto would find a new love, and Celia had charged you, Paula, with the same task, which you had not only fulfilled but in addition given us a wink from the Beyond: Giulia was born the same day as you—October 22. Her mother was named Paula and her father was born the same day and year that I was. Too many coincidences. I couldn't help but think that you had chosen her because she would make your husband happy.

Ernesto and Giulia hid their dismay concerning the disaster as best they could. Despite the dramatic circumstances we were living in, we had immediately given Giulia two thumbs up. She was perfect for Ernesto: strong, organized, happy, and affectionate.

According to Willie, we shouldn't worry about it, since that couple didn't need the approval of a family they had no blood ties with. "If they marry, we will have to bring them to California," I answered.

In the meantime, the turkey had turned green from the subcutaneous injection of herbs, and had come from the oven looking as poisonous as the air we were breathing in the house. Nico and Jason were undone, and could not take part in the wake—because that's what that day was, a wake. Alejandro and Nicole were sick in bed with a fever. Andrea was running around sucking her finger and dressed for the occasion in my sari, which she had wrapped herself in like a sausage. Willie was indignant because neither of his two sons had shown up. He was hungry, but no one had taken charge of the dinner, which on any normal Thanksgiving is a banquet. Following an uncontrollable impulse, my husband picked the green turkey up by one leg and threw it into the garbage.

Unfavorable Winds

The collapse of the family didn't happen overnight; for several months Nico, Celia, and Sally debated back and forth, but they never for a moment forgot the children. They tried to protect them as much as possible, despite the less than perfect situation they found themselves in. They took special pains to give the kids a lot of affection; in these dramas, however, suffering is inevitable. "It's all right, they will work it out later in therapy," Willie said to calm me. Celia and Nico kept living in the same house for a while because they had nowhere to go, while Sally came and went in her status of auntie. "This seems like a French film, I don't want to go over there any more," Tabra declared, scandalized. Even my tolerance didn't stretch that far; I decided not to go myself, but every day without seeing my grandchildren was like a funeral.

As I tried to keep close ties to Nico, who didn't give me much room, my relationship with Celia passed from tears and hugs to recriminations. She accused me of not understanding what was happening; I had a closed mind and I meddled in everything. Why the hell didn't I leave them in peace? I was offended by her explosive nature and brusque manners, but two hours later she would call and apologize and we would be reconciled . . . until the cycle was repeated. It was very painful for me to see her suffer. The decision she had made came at a very high cost, and all the passion in the world would not save her from paying it. Celia wondered if there weren't something perverse in her that led her to destroy the best of everything she had: her home, her children, a family in which she was safe, comfortable, cared for, loved. Her husband adored her and he was a very smart and very good man. Even so, she felt trapped in the relationship; she was bored, she didn't fit in her skin, her heart escaped in longings she couldn't name. She told me that the seemingly perfect edifice of her life had come tumbling down with Sally's first kiss. That was enough for her to realize that she could not go on living with Nico; in that one instant her destiny changed course. She knew that the reaction against her would be merciless, even in California, which prides itself on being the most liberal place on the planet.

"Do you think I'm abnormal, Isabel?" she asked me.

"No, Celia. A certain percentage of people are gay. The bad thing is that you came to that knowledge a little late, after you already had three children."

"I know that I'm going to lose all my friends and that my family will never talk to me again. My parents will never understand, you know the world I come from."

"If they can't accept you as you are, you can get along without them for the moment. There are other priorities right now; first of all, your children."

She stopped coming to my office because she didn't want to be dependent on me, she told me, but if she hadn't decided that, I would have had to. We couldn't go on working together. It was nearly impossible to replace her; I had to hire three people to do the work she had done alone. I was used to Celia, I had blind confidence in her, and she had learned to imitate me, from my signature to my style. We joked that one day in the not too distant future she'd be writing my books. Celia, Nico, and Sally began going to therapy, separately and together, to work out details. Celia was again prescribed antidepressants and sleeping pills; she was stupefied with medication.

As for Jason, no one thought much about him because he had decided to stay in New York after he graduated; there was nothing to attract him to California and he didn't want ever to see Sally or Celia again. He felt isolated; he had lost his family. He kept losing weight, and his looks changed. He was no longer a layabout kid, he'd turned into an angry man who spent a good part of the night wandering the streets of Manhattan because he couldn't sleep. There was no shortage of late-night girls he could tell his misfortunes to, who then consoled him in bed. "It would be three or four years before I trusted a woman again," he told me much later, when we could talk about it. He had also lost confidence in me because I hadn't gauged the degree of his suffering. "Stop acting like a pansy," Willie told him the first time he mentioned it; it was his favorite phrase for resolving his sons' emotional conflicts.

And me? I devoted myself to cooking and knitting. I got up at dawn every day, cooked quantities of food, and took some to Nico's house or left some on the roof of Celia's van, so at least they would have something to eat. I knit and knit an enormous, shapeless heavy wool article that according to Willie was a sweater for the house.

In the midst of this tragicomedy, my parents came for a visit and landed smack in the middle of one of those monstrous storms that are a blemish on the benign climate of northern California. It was as if nature wanted to illustrate our family's state of mind. My parents live in a pleasant apartment in a peaceful residential neighborhood in Santiago, among noble trees, where at dusk uniformed maids, still today in the twenty-first century, escort fragile old ladies and pampered dogs. They are looked after by Berta, who has worked for them for more than thirty years and is much more important in their lives than the seven children they have between them. Willie suggested once that they move to California and spend the rest of their days near us, but there isn't enough money to buy in the United States the comfort and company they enjoy in Chile. I am consoled by our separation when I think of my mother with her mustached painting instructor, her lady friends at Monday tea, sleeping her siesta between starched sheets, presiding at her table during banquets prepared by Berta, happy in her home filled with relatives and friends. Here old people are left to themselves. My mother and Tío Ramón come to see us at least once a year, and I go two or three times to Chile. In addition we have our daily contact via letters and telephone.

It is nearly impossible to hide anything from that pair of astute septuagenarians, but I had said nothing to them about the events with Celia. I was clinging to the vain illusion that with time the problem would resolve itself; perhaps it was only a caprice of the young. The result was that there was a notable void in my correspondence with my mother during those months, and to reconstruct this story I've had to question, separately, the participants and various witnesses. Each of them remembers things differently, but at least we can now talk about it openly. As soon as my parents set foot in San Francisco, they noticed that something very serious had rattled us, and we had no choice but to tell them the truth.

"Celia fell in love with Sally, Jason's fiancée." I just blurted it out.

"I hope no one knows this in Chile," my mother half-whispered when she could react.

"They will, you can't hide these things. Beside, it happens everywhere."

"Yes, but in Chile it isn't talked about."

"What are they going to do?" asked Tío Ramón.

"I don't know. The whole family is in therapy. An army of psychologists is getting rich off us."

"If we can help in any way . . . ," mother murmured—her love was always unconditional even when her voice was trembling—and added that we must let them work it out themselves. And be discreet, because our opinions would only aggravate matters.

"You start writing, Isabel, that will keep you occupied. And that way you won't be interfering more than is called for," Tío Ramón advised me.

"That's what Willie tells me."

But We Keep Paddling

M y Sisters of Disorder placed another candle on their altars, in addition to the ones they already had for Sabrina and Jennifer, and prayed for the rest of my unhinged family, and to get me back to my writing—I'd looked too long for excuses not to. The eighth of January was approaching and I hadn't been able to write a word of fiction; the discipline part I could manage, but I lacked the ease, although the trip to India had filled my head with images and color. I didn't feel paralyzed any longer; the well of inspiration was full, and more active than ever, because the idea of the foundation was beginning to take shape. However, to write a novel one needs crazed passion, which now was catching fire, but I needed to feed it oxygen and fuel to make it burn more brightly. I kept turning over in my mind the idea of "a memoir of the senses," an exploration on the theme of food and sexual love. Given the climate of the passions ruling my family, that might seem a little ironic, but that wasn't my intention. I had thought of it before the business of Celia and Sally. I even had a title, *Aphrodite*, which, being vague, gave me full liberty to go in any direction. My mother went with me to porno shops in San Francisco, seeking inspiration, and she offered to help with the part involving sensual cuisine. I asked her where she would find erotic recipes, and she replied that any dish presented flirtatiously is an aphrodisiac, so she wouldn't have the wasted energy of birds' nests and rhinoceros horn, so difficult to find in local markets. She had been brought up in one of the most Catholic and intolerant places in the world, and had never been in a shop "for adults," as they're called, and when I translated from English the instructions for various rubber appurtenances, she nearly choked from laughing. My research for "Aphrodite" triggered erotic dreams for both my mother and me. "At seventy-something, I still think about that," she confessed. I reminded her that my grandfather thought about it at ninety. Willie and Tío Ramón were our guinea pigs. We tested the aphrodisiac recipes on them; like black magic, they have effect only if the victim knows what he's been given. The same plate of oysters, without the explanation that it stimulates the libido, will have no visible results. Not everything was drama in those months, we also had some fun.

When we could, Tabra, my parents, and I went off to your forest to take walks. The rains fed the stream where we'd scattered your ashes, the woods smelled of pine and wet earth. We walked at a good pace, my mother and I in the lead, in silence, Tío Ramón and Tabra behind, talking about Che Guevara. My stepfather believes that Tabra is one of the most interesting and beautiful women he has ever known, and she admires him for many reasons, but none more than because he once met the heroic guerrilla. He was even photographed with him. Tío Ramón has repeated the same story two hundred times, but she never tires of hearing it nor he of telling it. You greeted us from the treetops; and joined us in our walk. I hadn't told my parents that once your ghost had come in a taxi to visit us at the house; I saw no reason to confuse them even more.

I have wondered where my tendency to live with spirits comes from; it seems that not everyone has this mania. First of all, I have to clarify that only rarely have I come face to face with a spirit, and on the occasions that has happened, I can't be sure I wasn't dreaming. I have no doubt, however, that your spirit is always with me. If not, why would I be writing these pages to you?

You manifest yourself in the strangest ways. For example, once when Nico was changing jobs it occurred to me to invent a corporation to hire him. I even got as far as consulting an accountant and a couple of attorneys who overwhelmed me with rules, laws, and figures. "If only I could call Paula and ask her advice!" I said aloud. At that moment the mail arrived and among the letters was an envelope for me, written in a hand so similar to mine that I opened it immediately. The letter consisted of a few lines written in pencil on a page of notebook paper. *"From now on I will not try to resolve others' problems until they ask me for help. I am not going to shoulder responsibilities that are not mine to bear. I am not going to be overly protective of Nico and my grandchildren."* The note was signed by me and bore a date seven months old. Then I remembered that I had gone to my grandchildren's school on Grandparents Day, and the teacher had asked everyone there to write a resolution or a wish and put it in an addressed envelope so that she could mail it later. There's nothing strange in that. What's strange is that it arrived at precisely the moment I was asking for help from you. Too many inexplicable things happen. The idea of spirit beings, real, imaginary, or metaphorical, originated with my maternal grandmother. That branch of the family has always been unconventional and has been a fertile source for my writing. I would never have written *The House of the Spirits* if my grandmother hadn't convinced me that the world is a very mysterious place.

The family situation evolved into a more or less normal pattern. Well, normal for California. In Chile it would have been a scandal worthy of the tabloids, especially because Celia found it necessary to announce the situation with a megaphone and preach the advantages of gay love. She said everyone should try it, that it was much better than being heterosexual, and she ridiculed men and their capricious "piripichos." I myself talked about it too much, so our plight flew from mouth to mouth and gossip swirled all around us. People we scarcely knew came up to express their sympathy, as if we were in mourning. I think all California knew. Much hullabaloo. At first I wanted to hide in a cave, but Willie convinced me that it isn't the truth exposed that makes us vulnerable, it's what we try to keep secret. Nico and Celia's divorce did not bring an end to things, because we were still mired in a swamp of constantly changing relationships, but we cut no ties. The three children bound us together, whether we wanted or not. Nico and Celia sold the house we had bought with such effort and divided the money.

They decided that the children would spend a week with their mother and the next with their father, that is, they would live with a suitcase on their backs, but that was preferable to the Solomonic solution of cutting them in half. Celia and Sally found a little house that needed repair, but it was very well located and they settled in as comfortably as they could. Things were very hard for them in the beginning; their families—people in general—turned their backs on them. They were on their own, with few resources and a sensation of having been judged and sentenced. I kept in close touch and tried to help, often behind Nico's back, as he could not understand my affection for the ex-daughter-in-law who had so wounded our family. Celia confessed to me that she cried nearly every day, and Sally had to swallow the gossipers' accusation that she had destroyed a family, but as the months went by, the noise started to fade, as nearly always happens.

Nico found an old house two blocks from ours, which he remodeled, rehabbing the floors, windows, and bathrooms. The house had a garden crowned by two enormous palm trees, and it looked out on the shore of a lagoon where geese and wild ducks nested.

He lived there with Celia's brother, to whom he had offered a place to stay for a year and who for some curious reason did not go with his own sister. This young man was still looking for his destiny, but without much success. It may have been because he didn't have a work permit, and his tourist visa, which had already been renewed a couple of times, was about to expire. He was often depressed or in a foul mood, and more than once Nico had to cut short the tantrums of a man who was no longer his brother-in-law but who continued to be his guest.

For Celia and Sally, who had jobs with flexible hours, taking care of the children in the week they came to them was not as complicated as it was for Nico, whose job was quite a distance away and who had to do it alone. Ligia, the same woman who had rocked Nicole during the months of her inconsolable crying, came to his aid and continued to do so for several years. She picked up my grandchildren at school, where there was a preschool for Andrea, a day care for Nicole, and a kindergarten for Alejandro, brought them to the house, and stayed with them until I arrived, if I could do it, or Nico, who tried to leave his office earlier during the week he had the children and make up the hours when he didn't. Nico never showed any sign of bewilderment or impatience; to the contrary, he was a happy, calm father. Thanks to his organization his household kept rolling, but he got up at dawn and went to bed very late, exhausted. "You don't have a minute for yourself,

Nico," I told him one day. "Yes, Mamá, I have two hours of quiet, all by myself, in the car as I drive to and from the office. The more traffic the better," he replied.

Nico and Celia's relations turned bitter, and Nico defended his territory as best he could; the truth of the matter is that I didn't help him in that unpleasant task. Finally, tired of gossip and small betrayals, he asked me to end my friendship with Celia because as it was, he was having to fight on two fronts. He felt ignored and impotent as the children's coparent, and trampled on by his own mother. Celia came to me if she needed something, and I never consulted Nico before I acted, so I sabotaged some of the decisions they'd previously agreed on but Celia had later changed. In addition, I lied to avoid explanations, and of course, I was always caught out. For example, the children took it upon themselves to tell Nico when they had seen me the day before at their mother's house.

Abuela Hilda, perplexed over the course of events, went back to Chile to stay with her only daughter. We never heard a single word of criticism from her. She refrained from giving her opinion, faithful to her formula for averting conflicts, but her daughter, Hildita, told me that she was taking one of her mysterious green pills for happiness every three hours. They had a magical effect, because when a year later she came back to California, she was able to visit Celia and Sally with her old affection. "These girls are such good friends; it's a pleasure to see how they get along," she said, repeating a comment she'd made many years before, when no one suspected what was going to happen.

A Tribe in Distress

E arly on, I hid in the bathroom to call and set up clandestine meetings with Celia. Willie heard me whispering and began to suspect that I had a lover. Nothing more flattering, because one look at my body and he would understand that I would never undress in front of anyone but him. In truth, however, my husband wouldn't have the strength for a fit of jealousy. During that period he had more cases than ever, and still refused to give up on the matter of Jovito Pacheco, the Mexican who'd fallen from the scaffold of a building under construction in San Francisco. When the insurance company refused to make payment, Willie filed suit. The jury selection was critical, as he explained to me, because there was a growing hostility against Latino immigrants and it was nearly impossible to get a sympathetic jury. In his long experience as a lawyer, he'd learned to excuse obese persons from the jury—for some reason they always voted against him—but now he also had to excuse racists and xenophobes, who were more numerous every day. In California the hostility between Anglos and Mexicans is very old, but a piece of legislation—Proposition 187—truly lit a fuse under that sentiment. North Americans love the idea of immigration; it's basic to the American dream that the son of some poor devil who comes to these shores carrying a cardboard suitcase can become a millionaire; but they detest the immigrants. That hatred, which marked Scandinavians, Irish, Italians, Jews, Arabs, and many others, is worse when directed toward people of color, and especially toward Hispanics, I suppose because there are so many of them, and there is no way to keep them from coming.

Willie traveled to Mexico, rented a car, and by following the complicated directions he'd been given in a letter, drove three days, snaking along dusty tracks to reach a remote village of adobe houses. He was carrying a yellowing photograph of the Pacheco family, which he used to identify his clients: a grandmother of iron, a timid widow, and four fatherless children, one of them blind. They had never worn shoes, they had no potable water or electricity, and they slept on straw mats on the ground. Willie convinced the grandmother, who directed the family with a firm hand, that they should come to California to be present at the trial, and he assured her that he would send her funds to do it. When he left to return to Mexico City, he found that the main highway passed the little village only five hundred meters away; none of his clients had ever used it, which was why they had sent instructions to follow the mule trails. He made the return to the city in four hours. He made arrangements to obtain visas for the Pachecos' brief visit to the United States, got them on a plane, and brought them there—mute with fear at the idea of going up in a big tin bird. Once they were in San Francisco, he discovered that the family was not comfortable in any motel, however modest; they knew nothing about plates or silverware—they ate using tortillas—and they had never seen a toilet. Willie had to demonstrate for them, which produced attacks of giggles from the children and perplexity in the two women. They were intimidated by that enormous city of cement, the streams of traffic, and all the people speaking some incomprehensible gibberish. Finally he put them up with another Mexican family. The children settled in front of the television, incredulous before such a miracle, while Willie tried to explain to the grandmother and the widow what a trial consisted of in the United States.

On the appointed day he appeared in court with the Pachecos, the grandmother in the lead, wrapped in her rebozo, wearing flip-flops she could barely keep on her wide campesina's feet, and understanding not one word of English, and behind her came the widow and the children. In his closing argument, Willie coined a phrase that we have teased him about for years: "Ladies and gentlemen of the jury, are you going to allow the lawyer for the defense to toss this poor family onto the garbage heap of history?" Not even that convinced them. They gave nothing to the Pachecos. "This would never have happened to a white person," Willie commented, as he prepared an appeal before a superior court. He was indignant about the jury's verdict, but the family took it with the indifference of people accustomed to misfortune. They expected very little from life and did not understand why that lawyer with blue eyes had gone to all the trouble of coming to look for them in their village in order to show them how a toilet worked.

To ease the frustration of having failed them, Willie decided to take them to Disneyland in Los Angeles, so at least they would have a good memory from the trip.

"Why are you creating expectations for those children that they will never satisfy?" I asked him.

"They need to know what the world offers, so they can prosper. I made it out of the wretched ghetto where

I was raised because I realized that I could aspire to more," was his answer.

"You are a white male, Willie. And as you yourself say, whites have the advantage."

My grandchildren got used to the routine of changing homes every week, and of seeing their mother as a couple with Aunt Sally. It was not an unheard-of arrangement in California, where domestic relationships are very flexible. Celia and Nico went to the children's school to explain what had happened, and the teachers told them not to worry; by the time the students reached the fourth grade, 80 percent of their classmates would have stepmothers or stepfathers, and often there would be three of the same sex; they would have adopted brothers and sisters of other races, or they would be living with grandparents. The storybook family no longer existed.

Sally had seen the children being born and she loved them so much that years later, when I asked if she didn't want to have children of her own, she answered, "Why? I already have three." She assumed the role of mother with open heart, something I had never been able to do with my stepchildren, and for that alone I have always respected her. Nonetheless, I was once wicked enough to accuse her of having seduced half of my family. How could I have said something so stupid? She wasn't the siren enticing victims to crash on the rocks; everyone involved was responsible for his or her own acts and feelings. Besides, I had no moral authority to judge anyone; during my lifetime I'd done several insane things for love, and who knows whether I may do more before I die. Love is a lightning bolt that strikes suddenly, changing us. That is what happened to me with Willie, so why wouldn't I understand the love between Celia and Sally.

I received a letter from Celia's mother in which she accused me of having perverted her daughter with my satanic ideas, and "having stained her beautiful family, in which an error was always called an error, and a sin, a sin," very different from what I disseminated in my books and my conduct. I suppose that she couldn't imagine that Celia could be gay and that her daughter's problem was that she didn't know it; she married and had three children before she could admit it. What motive could I possibly have to induce my daughter-in-law to wound my family? It seemed extraordinary to me that someone would attribute such power to me.

"What luck! Now we never have to speak to that woman again," were Willie's first words when he read the letter.

"Seen from outside, Willie, we may give the impression of being very decadent."

"You can't know what happens in other families behind their closed doors. The difference with ours is that everything is out in broad daylight."

I was feeling calmer in regard to my grandchildren. I was counting on the dedication of their parents; they had more or less the same rules in both houses, and the school offered stability. The children were not going to end up traumatized but, rather, overly indulged. They had been given such honest explanations that sometimes they chose not to ask because the answer might go further than they wanted to hear. From the beginning, I established the practice of seeing them almost every day when they were with Nico, and a couple of times a week at Celia and Sally's house. Nico was firm and consistent; his rules were clear and at the same time he lavished tenderness and patience on his children. Many Sundays I surprised him early in the morning asleep with all of them in his bed, and nothing moved me so much as to see him come in the door with the two girls in his arms and Alejandro clinging to his legs. In Celia's house there was a relaxed atmosphere, clutter, music, and two skittish cats that shed on all the furniture. They often improvised a tent with quilts in the living room, where the children would camp the whole week. I think Sally was the one who kept the seams of that family from ripping apart; without her I think Celia would have gone down in that period of such high stress. Sally had a sure hand with the children; she sensed problems before they happened, and kept a close watch over them without smothering them.

I reserved "special days" with each child, separately; on that day they got to choose the activity. That is why I had to sit through the animated version of *Tarzan* thirteen times, and one called *Mulan* seventeen. I could recite the dialogue backward. They always wanted the same thing on their special day: pizza, ice cream, and a movie, except once when Alejandro showed interest in seeing the men dressed as nuns who'd been on the television. A group of homosexuals, theater people, made themselves up as nuns, painted their faces, and paraded around collecting money for charity. The folly of this enterprise was that they did it during Easter week. It was on the news because the Catholic Church had ordered its parishioners not to visit San Francisco, hoping to cripple tourism in a city that, like Sodom and Gomorrah, lived in mortal sin. I took Alejandro to see *Tarzan* one more time.

Nico had become very quiet, and there was a new hardness in his eyes. Rage had closed him up like an oyster; he didn't share his feelings with anyone. He wasn't the only one who suffered, each of us had some part of it, but he and Jason stood alone. I clung to the consolation that no one had acted maliciously, it was simply one of those storms in which the ship's wheel spins out of control. What had happened between Celia and him behind closed doors? What role did Sally play? It was futile to try to sound him out; he always answered with a kiss on my forehead and some unconnected comment to distract me, but I have not lost the hope that I will find out in my final hour, when he will not dare refuse the wish of his dying mother. Nico's life was reduced to work and looking after his children. He had never been very sociable; Celia was the one responsible for their friends, and he had made no effort to keep in touch with them. He had isolated himself.

While all this was going on, a psychiatrist who had the looks of a movie actor and aspirations of a novelist came to wash our windows; he earned more doing that than he did listening to the tiring platitudes of his patients. In truth, he didn't do the actual work, that was done by two or three splendid Dutch girls. I have no idea where he found them; they were always different, but all were bronzed by California sun, with platinum hair and short shorts. These beauties climbed ladders with rags and pails while he sat in the kitchen and told me the plot of his next novel. It made me angry, not just for the dumb blondes who did the heavy work, which he was paid for, but that this man, not even the shadow of Nico, had all the women he wanted. I asked him how he did it, and he said, "I just lend an ear; women want to be heard." I decided to pass that information along to Nico. Even with his arrogance, the psychiatrist was easier to take than the old hippie who had preceded him in the window-cleaning department. Before he would accept a cup of tea, he painstakingly checked out the teapot to be sure that it was free of lead; he talked in whispers, and once spent fifteen minutes trying to get an insect off the window without hurting it. He nearly fell off the ladder when I offered him a fly swatter.

I was keeping close tabs on Nico, and we saw each other nearly every day, but he had become a stranger to me, every day more reserved and distant, although his impeccable courtesy never deserted him. Such delicacy came to irritate me, I would have preferred a little hair pulling. After two or three months, I couldn't stand it any longer and I decided that we couldn't keep putting off a really frank conversation. Confrontations are very rare between us, partly because we get along fine without a great show of sentiment and partly because that's how we are by nature and by habit. Through the twenty-five years of my first marriage, no one ever raised his voice; my children grew up with an absurd British urbanity. Furthermore, we start from good intentions, and if someone is offended, it happens by error or omission, not any spirit of wounding the other. For the first time I blackmailed my son. In a quivering voice I reminded him of my unconditional love, and of all the things I had done for him and his children from the day they were born; I reproached him for withdrawing and keeping everything to himself . . . in sum, a pathetic speech. I have to admit that he has always been a prince with me—with the exception of the time he was twelve and played a nasty trick on me by pretending he had hanged himself. I know you remember the time your brother strung up a harness in a door frame. When I saw him with his tongue protruding and a thick rope around his neck I very nearly departed this world. I will never forgive him for that! "Why don't we just get to the point, Mamá?" he asked amiably, after listening quite a while, unable to hide any longer that he was gazing at the ceiling with boredom. At that point, we launched into full attack. We came to a civilized agreement: he would make an effort to be more present in my life, and I would make an effort to be more absent in his. That is, neither bald or with two wigs, as they say in Venezuela. I had no intention

of carrying out my part of the deal, as he saw immediately when I suggested that he try to meet women
because at his age it's not good to be celibate: you have
to use it or lose it.

"I heard that at one of your office parties you were
talking with a very nice girl. Who is she?" I asked.

"How did you know that?" he answered with
alarm.

"I have my sources of information. Are you thinking of calling her?"

"I have all I need with three children, Mamá. I don't
have time for romance," and he laughed.

I was sure that Nico could attract any woman he
pleased; he had the looks of an Italian Renaissance nobleman. He had a good disposition—he got that from
his father—and he wasn't stupid—he got that from
me—but if he didn't get in gear he was going to end
up in a Trappist monastery. I told him about the psychiatrist with his court of Dutch girls who washed our
windows, but he didn't evince the least interest. "Keep
your nose out of it," Willie said, as he always does. Of
course I was going to stick my nose in, but first I would
have to give Nico a little time to lick his wounds.

PART TWO

The Onset of Autumn

According to the dictionary, autumn is not only the golden season of the year, but also the age when we cease to be young. Willie would soon be sixty and I was striding firmly through the decade of my fifties, but my youth ended at your side, Paula, in the corridor of lost steps in that Madrid hospital. I felt my maturing as a journey inward and the beginning of a new kind of freedom: I could wear comfortable shoes and I no longer had to diet or please half the world, only those who truly mattered to me. I had always had my antennae extended to capture the male energy in the air; after fifty those antennae began to droop and now only Willie attracts me. Well, maybe Antonio Banderas too, but that's purely hypothetical. Willie and I aged, our bodies and minds changed. His prodigious memory began to stutter a little, he no longer remembers the phone numbers of all his friends and acquaintances. His shoulders and knees grew stiff, his allergies grew worse, and I got used to hearing him coughing every minute like an old locomotive. As for him, he resigned himself to my peculiarities: emotional problems tie my stomach in knots and give me headaches; I can't watch bloody movies, I don't enjoy parties, I eat dark chocolate in secret, I fly off the handle, and I spend money as if it grew on trees. In this onset of autumn we finally came to know each other and accept each other uncondition-ally; our relationship grew richer. Being together is as natural as breathing, and sexual passion turned into calmer and more tender moments. But chastity? No. We're bound together, we don't want to be separated, but that doesn't mean we don't have an occasional fight. I never lay down my sword. Just in case.

On one of our trips to New York, an obligatory stop on all the tours for promoting my books, we visited Ernesto and Giulia in their home in New Jersey. When they opened the door, the first thing we saw as we went in was a small altar holding a cross, Ernesto's aikido weapons, a candle, two roses in a vase, and a photograph of you. The house had the same air of whiteness and simplicity as the spaces you had decorated during your brief life, perhaps because Ernesto shared the same tastes. "She protects us," Giulia told us, gesturing toward your photo as we passed, in a completely natural voice. I realized that this young woman had been intelligent enough to adopt you as a friend instead of competing with your memory, and in that way she had gained the admiration of Ernesto's family, who adored you, and of course ours. Right then I began to plan how I could get the couple settled in California, where they could be part of the tribe. But what tribe? There wasn't much left: Jason in New York, Celia with a new partner, Nico angry and absorbed by the kids, my three grandchildren going and coming with their little clown suitcases, my parents in Chile, and Tabra traveling the unknown corners of the world. Even Sabrina was going to preschool; she had her own life and we seldom saw her. She could now get around with a walker, and had asked for a bigger bicycle for Christmas.

"We're running out of tribe, Willie. We have to do something soon or we'll end up playing bingo in some geriatric retirement community in Florida, like so many American senior citizens who might as well be living on the moon."

"And what is the alternative?" my husband asked, undoubtedly thinking of death.

"Be a burden to the family, but to do that, we have to add to it," I informed him.

I was joking, of course, because the worst thing about old age isn't loneliness but being dependent. I don't want to inflict my decrepitude on my son and grandchildren, though it wouldn't be bad to spend my last years near them. I made a list of priorities for my eighties: health, financial resources, family, dog, stories. The first two would allow me to decide how and where to live, the third and fourth would keep me company, and the stories would keep me quiet and entertained and not driving anyone crazy. Willie and I are both terrified of losing our minds, in which case Nico, or even worse, strangers, would have to take charge of us. I think of you, Paula, spending months at the mercy of other people before we could bring you to California. How many times had you been mistreated by a doctor, a nurse, or an employee and I didn't know about it? How many times had you wished in the silence of that year to die soon, and in peace?

The years slip by, stealthily, on tiptoes; they whisper behind our back, making fun of us, then suddenly one day they frighten us when we look in the mirror, they drop us to our knees or drive a dagger into our backs. Old age attacks us every day, but it seems to be most evident at the end of every decade. I have a photo of me at forty-nine, at the launch for *The Infinite Plan* in Spain. It's the picture of a young woman, hands on her hips, defiant, with a red shawl thrown around her shoulders, her fingernails painted, and wearing Tabra's long earrings. It was that same moment, with Antonio Banderas at my side and a glass of champagne in my hands, that they came to tell me that you were in the hospital. I ran out of there, never imagining that your life and my youth would be coming to an end. Another of my photos, a year later, shows a mature woman, hair cut short, eyes sad, clothing dark, no adornment. My body had become a burden; I looked at myself in the mirror and didn't recognize myself. However, it was not only sorrow that suddenly aged me; when I look through the family photo albums I can see that when I turned thirty, and later forty, there was also a drastic change in my appearance. It will be that way in the future, except that instead of noticing at the end of a decade, it will be with every leap year, my mother tells me. She is twenty years ahead of me, leading the way, showing me how it will be at every stage of my life. "Take calcium and hormones so your bones don't get brittle, like mine," she advises me. She repeatedly tells me to pamper myself, savor the hours because they go very quickly; I should never stop writing, it keeps my mind active, and I should do yoga so I can bend over and put my shoes on by myself. She adds not to work too hard trying to look young because your years show no matter what, however much you try to disguise them, and there's nothing as ridiculous as an old woman done up like Lolita. There are no magic tricks to prevent deterioration, you can only postpone it a little. "After turning fifty, vanity becomes equivalent to suffering," says this woman with a reputation for being a beauty. But I fear the ugliness of old age, and I plan to fight it as long as I'm healthy, which is why I had cosmetic surgery, since the snake oil that will rejuvenate cells has yet to be discovered. I wasn't born with the splendid raw material of a Sophia Loren, I need all the help I can get. The operation is like detaching muscles and skin, cutting away the excess, and sewing the flesh back to the skull, snug as a ballerina's tights. For weeks I had the sensation that I was wearing a wood mask, but in the end it was worth the pain. A good surgeon can trick time. This is a subject I can't discuss with my Sisters of Disorder, or with Nico, because they believe that old age has its own beauty, including varicose veins and those warts with hairs. You agree with them. You always preferred old people to children.

In Bad Hands

On the subject of plastic surgery, one early Wednesday morning Tabra called, somewhat disturbed, with the news that one of her breasts had disappeared.

"Is this a joke?"

"It went flat. One side is smooth, but the other breast is like new. Nothing hurts. Do you think I should see a doctor?"

I immediately picked her up and took her to the surgeon who'd done the procedure. He assured us it wasn't his doing, the fault lay with the implant manufacturer; sometimes they are defective, they burst, and the fluid spreads through the body. It wasn't anything to worry about, he added, it's a saline solution that over time is absorbed with no danger to your health. "But she can't go around with one breast!" I intervened. That seemed reasonable to him, and a few days later he replaced the punctured implant, although it didn't occur to him to give her a discount on the price of his services. Three weeks later, the second breast deflated. Tabra came to our house wearing a poncho.

"If that bastard doesn't take responsibility for your tits, I'll sue him!" Willie bellowed. "I'll see that he replaces that one for nothing!"

"I really don't want to bother him again, Willie. He might get angry. I went to see a different doctor," she admitted.

"And does this one know anything about breasts?" I asked.

"This is a very decent man. He goes to Nicaragua every year to operate on children who have a harelip. Free of charge."

In fact, he did an excellent job, and Tabra will have the firm breasts of a damsel until she dies at the age of one hundred. The women in her family live very long. Within a few months the first surgeon—he of the failed implants—appeared in the news. He had operated on a patient and left her in his office overnight without a nurse, and the woman had suffered an attack and died. My grandson Alejandro figured the cost of each of his aunt Tabra's breasts and suggested that she could charge ten dollars for looking and fifteen for touching, and that way recoup her investment in about three years, one hundred and fifty days. Tabra, however, was doing fine with her jewelry and did not have to take such extreme measures.

In view of her booming business, Tabra hired a manager full of grandiose ideas. She had created her business from nothing; she had begun selling her pieces in the street, and step by step, with hard work, perseverance, and talent, she had built a model enterprise. I couldn't understand why she needed a man who had never crafted a bracelet in his life, nor worn one. He couldn't even boast of black hair. And she knew much more than he did. This business school graduate had a friend in the computer business, and he began by buying a bank of computers that equaled NASA's; none of Tabra's Asian refugees learned to use them even though several of them spoke four or five languages and were well educated. He then decided that what was needed was a group of consultants to form a board of directors. Those he selected from among his friends, to whom he paid a handsome salary. In less than a year, Tabra's business was staggering along like Willie's office; more money was going out than coming in, and she had to keep an army of employees whose functions no one understood. These expenses coincided with a time in which the economy was taking a dive, minimalist jewelry was in vogue, not Tabra's large ethnic pieces, and the company was badly administered. That was the moment this financial whiz chose to make a change, leaving Tabra drowning in debts. He was hired as a consultant in other businesses, recommended by the very people he'd hired for Tabra's board.

For months Tabra struggled with creditors and pressure from the banks, but in the end she had to resign herself and declare bankruptcy. She lost everything. She sold her idyllic property in the woods for much less than she had paid for it. Her belongings were appropriated, from her van to her factory equipment, and most of the inventory she'd acquired during a lifetime. Months before, Tabra had given me vials of beads and semiprecious stones that I kept in our cellar, waiting for the moment that she would have time to teach me how to make a few necklaces, never suspecting that later they would help her get back to work. Willie and I emptied and painted the room that had been yours and offered it to her so she would at least have a family and a roof over her head. She moved with the little bit of furniture and art she could save. We provided her with a large table, and there she began again, crafting her pieces one by one, as she had thirty years before.

We went out almost every day to walk and talk about life. I never heard her complain or curse the man who'd ruined her. "It's my fault for hiring him, and it will never happen to me again," was all she said. In the years that I've known Tabra, which add up to a lot, my friend has been sick, disillusioned, poor, and with a thousand problems, but I have seen her despair only once: when her father died. She cried for months over that man she adored and for other losses in the past, and I could not console her. In the period of her financial travails her demeanor never changed. With humor and courage she prepared to travel from the beginning the road she'd traveled in her youth, convinced that if she had done it at twenty she could do it again at fifty. She had the advantage of having a name recognized in several countries; anyone in the ethnic jewelry trade knows who Tabra is. Owners of art galleries come to her from Japan, England, the Caribbean islands, and she has clients who collect her work obsessively; they have more than five hundred pieces and keep buying.

Tabra proved to be an ideal guest. Out of courtesy she ate whatever was on her plate, and without our daily walks she would have ended up round. She was discreet, silent, and good company, and in addition she entertained us with her opinions.

"Whales are misogynists because when the female is in rut the males surround her and rape her," she told us.

"You can't apply the criteria of Christian morality to cetaceans," Willie rebutted.

"Morality is morality, Willie."

"The Yanomamo Indians in the Amazon jungle rape women of other tribes, and they're also polygamous."

Then Tabra, who has a great respect for primitive peoples, concluded that you can't, in fact, apply the same standards of morality to Yanomamos and to whales. And that isn't even a shadow of the political discussions! Willie is very liberal, but compared to Tabra he belongs among the Taliban.

To entertain herself following another of the sudden disappearances of Alfredo López Lagarto-Emplumado—this one coinciding with her bankruptcy—Tabra returned to the vice of arranging blind dates through ads in the newspapers. One of the candidates presented himself with his shirt open to the belly button and sporting a half dozen gold crosses on his hairy chest. That, more than the fact that he was white and getting bald on the crown of his head, should have been enough to squelch my friend's interest, but he seemed intelligent and she decided to give him a chance. They met at a cafeteria, sat talking for a long while, and discovered they had interests in common, like Che Guevara and other heroic guerrillas. On the second date, the man had buttoned up his shirt, and he also brought her a beautifully wrapped gift. When she opened it, she found a penis of optimistic dimensions carved in wood. Tabra came home in a rage, and threw it into the fireplace, but Willie convinced her that it was an objet d'art, and if she collected gourds that covered male privates in New Guinea, he saw no reason for her to be offended by that present. Even with her doubts, she went out once more with her gallant. On this third date, they ran out of topics related to Latin American guerrillas and sat for some time in silence, until Tabra, to say *something*, told him that she liked tomatoes. "I like *your* tomatoes," he replied, and grabbed the breast that had cost her so much money. And since she was paralyzed with outrage, he interpreted her stupor as authorization to take the next step, and invited her to an orgy in which the participants shed their clothes and dived headfirst into a pit of living flesh, to frolic like Romans in the times of Nero. Apparently a California custom. Tabra blamed Willie. She said that the penis had not been an artistic gift but a dishonest proposition and an assault on her decency, as she had suspected. There were other suitors. Very entertaining for us, though much less so for her.

Tabra was not the only person to furnish us with surprises. We learned about then that Sally was planning to marry Celia's brother, I suppose to provide him with a visa that would allow him to stay in the States. To convince the Immigration Service that it would be a legitimate marriage, they had a party, with a cake, and they took a photo in which Sally is wearing the famous wedding gown that had languished in my closet for years. I begged Celia to hide the picture because there was no way to explain to the children that their mother's partner was going to marry their uncle, but Celia doesn't like secrets. She says that everything comes out in the long run and there is nothing more dangerous than lies. And that was precisely why in the end the wedding never took place.

Searching for a Bride

Nico had become very handsome. He was wearing his hair long, like an apostle, and his grandfather's features had become more accentuated: large sultry eyes, aristocratic nose, square chin, elegant hands. It was inexplicable to me that there weren't a dozen women milling about at his front door. Behind Willie's back—he doesn't understand these matters—Tabra and I decided to look for a girlfriend for Nico. And that's exactly what you would have done, daughter, so don't scold me.

"In India, and many other places in the world, marriages are arranged. There are fewer divorces there than in Western countries," Tabra explained.

"That doesn't prove that they're happy, only that they have to put up with more," I contended.

"The system works fine. Marrying for love carries a lot of problems with it, it's more successful to unite two compatible persons who with time will learn to love one another."

"That's a little risky, but I don't have a better idea," I admitted.

It isn't easy to make these arrangements in California, as she herself had proved for years; none of the match-making agencies had found a man who was worth her while. The best had been Lagarto-Emplumado, but still she had no news of him. We checked the newspapers regularly to see if Moctezuma's crown had been returned to Mexico, but found nothing. In view of the negative results obtained by Tabra, I didn't want to put ads in the papers or go to agencies; that seemed a little indiscreet in view of the fact that I hadn't as yet consulted Nico. My friends were no help; they were no longer young, and no menopausal woman would take on my three grandchildren, however gorgeous Nico was.

I devoted myself to looking for a potential sweetheart everywhere I went, and in the process my eye grew sharper. I made inquiries among people I knew, I scrutinized the young women who asked for my autograph in bookstores, I even brazenly stopped a pair of girls in the street, but that method was inefficient and very slow. At that pace Nico would be seventy and still a bachelor.

I studied women, and in the end would discard them for different motives: serious or tedious, talkative or shy, smokers or macrobiotic fiends, dressed like their mothers or with a tattoo of the Virgin of Guadalupe on their backs. This was for my son; the choice could not be made frivolously. I was beginning to lose hope when Tabra introduced me to Amanda, a photographer and writer who wanted to make a trip to the Amazon with me for a travel magazine. Amanda was very interesting, and beautiful, but she was married and planned to have children very soon; she wasn't, unfortunately, a good candidate for my romantic designs. However, during one conversation, the subject of my son came up and I told her the whole drama—there was no secret about what had happened with Celia; she herself had broadcast it right and left. Amanda told me she knew the ideal girl: Lori Barra. Lori was her best friend; she had a generous heart, she had no children, she was pretty, refined, a graphic designer from New York who now lived in San Francisco. She had an obnoxious boyfriend, according to Amanda, but we'd find a way to get rid of him and leave Lori available to meet Nico. Not so fast, I said. First I need to know this girl through and through. Amanda organized a lunch, and I took Andrea with me; it seemed to me that at least the young designer ought to have a vague idea of what she would be taking on. Of the three children, Andrea was without doubt the most peculiar. My granddaughter came dressed like a beggar, with pink rags tied around different parts of her body, a straw hat with faded flowers, and her Save-the-Tuna doll. I was on the verge of dragging her somewhere to buy a more presentable outfit, but I decided it was best for Lori to know her in her natural state.

Amanda had said nothing to her friend about our plans, nor I to Nico; we didn't want to alarm them. The lunch in the Japanese restaurant was a good strategy; it didn't raise Lori's suspicion; she wanted to meet us only because she loved Tabra's jewelry and she had read a couple of my books: two points in her favor. Tabra and I were very impressed with her; she was a calm pool of simplicity and charm. Andrea observed her without saying a word, as she tried in vain to get pieces of raw fish into her mouth with chopsticks.

"You don't get to know a person in one hour," Tabra warned me afterward.

"She's perfect! She even looks like Nico. They're both tall, slim, handsome, have noble bones, and they wear black. They look like twins."

"Looking like twins isn't the basis of a good marriage."

"In India they have horoscopes, and let's say that isn't very scientific either. It's all a question of luck, Tabra," I answered.

"We need to know more about her. We have to see her in difficult circumstances."

"You mean like in a war?"

"That would be ideal, but they're all pretty far away. What do you say we invite her to go with us to the Amazon?" was Tabra's suggestion.

And that was how Lori, who had seen us only once, over a plate of sushi, ended up flying with us to Brazil in the role of assistant to Amanda.

When we planned the odyssey to the Amazon, I had imagined that we'd be going to a very primitive place where the character of Lori and others in the expedition would stand out clearly, but unfortunately the trip turned out to be much less dangerous than I'd expected. Amanda had seen to every last detail, and we reached Manaus without a hitch. We stopped for a few days in Bahía to meet Jorge Amado; Tabra and I had read all of his books and I wanted to know if the man was as extraordinary as the writer.

Jorge Amado and his wife, Zélia Gattai, received us in their home; he was seated in a large easy chair, amiable and hospitable. At eighty-four, half blind and not very well, he still had the sense of humor and the intelligence that characterize his novels. He was the spiritual father of Bahía. There were quotations from his books everywhere: chiseled in stone, adorning the facades of municipal buildings, in graffiti and primitive paintings on the huts of the poor. Plazas and streets proudly bore the names of his books and his characters. Amado invited us to try the culinary delights of his land in the restaurant run by Dadá, a beautiful black woman who was not the inspiration for his famous novel *Doña Flor and Her Two Husbands*—she was a child when he wrote it—but fit the description of the character: pretty, small, and agreeably plump without being fat. This replica of Doña Flor regaled us with more than twenty succulent dishes and a sampling of her desserts, which ended with little cakes of *punhetinha*, which in the local slang means "masturbation." Needless to say, all of this was very helpful for my *Aphrodite*.

The elderly writer also took us to a *terreiro*, or temple, in which he served as protective father, to witness a ceremony of Candomblé, a religion African slaves brought to Brazil several centuries ago, and that today has more than two million followers in that country, including urban middle-class whites. The divine rituals had started earlier with the sacrifice of some animals to the gods—orishas—but we didn't witness that part. The ceremony took place in a building that looked like a modest school, decorated with crepe paper and photographs of the *maes*, or "mothers," now dead. We sat on hard wooden benches and soon musicians arrived and began to beat their drums in an irresistible rhythm. A long line of women dressed in white entered and whirled with upheld arms around a sacred pillar, summoning the orishas. One by one they fell in a trance. No foaming at the mouth or violent convulsions, no black candles or serpents, no terrifying masks or bloody rooster heads. The older women carried to another room those who had been "mounted" by the gods and then brought them back adorned with the colorful attributes of their orishas, to keep dancing till dawn, when the liturgy concluded with an abundant meal of the roasted meat of the sacrificed animals, cassava, and sweets.

It was explained to me that each person belongs to an orisha—sometimes more than one—and at any moment of life you may be claimed and have to put yourself at the service of your deity. I wanted to know who mine was. One *mae de santo*, an enormous woman dressed in a tent of ruffles and lace, wearing a turban made from several kerchiefs and a profusion of necklaces and bracelets, "cast the shells" for us; there it's called *jogo de búzios*. I pushed Lori forward to get her reading first and the shells announced a cryptic new love: "Someone she knew but hadn't yet seen." Tabra and I had talked a lot about Nico, trying, of course, not to reveal our intention, and if by then Lori didn't know him it was because she had been on the moon. Will I have children? Lori asked. Three, the shells replied. Aha! I exclaimed, enchanted, but one look from Tabra brought me back to my senses. Then it was my turn. The *mae de santo* rubbed a handful of little shells between her palms for a long time, had me feel them in mine, and then threw them onto a black cloth. "You belong to Yemayá, the goddess of the oceans, mother of all things. Life begins with Yemayá. She is strong, a protector, she cares for her children, comforts them, and helps them in their sorrow. She can cure infertility in women. Yemayá is compassionate, but when she is angry she is terrible, like a storm on the ocean." She added that I had gone through great suffering, and that it had paralyzed me for a time but was beginning to dissipate. Tabra, who does not believe in these things, had to admit that at least the part about being maternal fit me. "She hit that accidentally," was her conclusion.

Seen from the plane, the Amazon is green as far as you can see. Below us is a mysterious land of water:

vapor, rain, rivers wide as seas, sweat. The Amazon territory occupies sixty per cent of the surface of Brazil, an area larger than India, and it also forms part of Venezuela, Colombia, and Peru. In some regions the "law of the jungle" still rules among bandits and traffickers in drugs, gold, wood, and animals who kill each other and, if they cannot exterminate the Indians with impunity, drive them from their lands. It is a continent in itself, a mysterious and fascinating world. It seemed so incomprehensible in its immensity that I couldn't imagine how it could serve me as inspiration, but several years later I would use a lot of what I saw there in my first novel for young readers.

To summarize the trip, since the details aren't relevant to what I'm relating, I can say that it was much safer than I'd wished. We'd prepared for a dramatic adventure in the world of Tarzan, but the closest link was a flea-bitten black female monkey who latched onto me and waited at my door from the break of dawn to climb onto my shoulders, curl her tail around my neck, and comb my head for fleas with her elfin fingers. It was a delicate romance. The rest was an eco-tourism stroll; mosquitoes were bearable, the piranhas did not tear out pieces of our flesh, and we did not have to dodge poisoned arrows; smugglers, soldiers, bandits, and traffickers passed by without seeing us; we did not get malaria, and worms did not burrow beneath our skin or fish like needles invade our urinary tracts. We four adventurers got away safe and sound. Nonetheless, this little adventure fully fitted our purposes: I got to know Lori.

Five Bullets

Lori passed with high marks. She was just as Amanda had described her: a clear mind and natural goodness. Discreetly and efficiently, she lightened the load for the rest of us, resolved annoying details, and soothed inevitable frictions. She had good manners, fundamental for sane coexistence, long legs, never a negative, and a frank smile that undoubtedly would seduce Nico. She had the advantage of being a few years older than he, since experience is always helpful, but she looked very young. She was striking, with strong features, a stupendous head of dark, curly hair, and golden eyes. None of that was relevant since my son doesn't place any importance on physical appearance; he scolds me because I use makeup and does not want to believe that when my face is washed clean I look like a rifleman. I observed Lori the way a hyena observes its prey, and even set a few traps for her, but I could not catch her in any fault. That made me a little uneasy.

After a couple of weeks, exhausted, we returned to Rio de Janeiro, where we were to take a plane to California. We stayed at a hotel in Copacabana, and instead of sunning on the white sand beaches, it occurred to us to go visit a *favela*, to get an idea of how the poor live, and to look for another seer to cast the *jogo de búzios*, since Tabra kept annoying me with her skepticism about my goddess Yemayá. We went with a female Brazilian journalist and a driver, who took us up to the hills of the unimaginably poor, an area where the police never went, to say nothing of tourists. In a *terreiro* much more modest than the one in Bahía, we were greeted by a middle-aged woman wearing jeans. The priestess repeated the ritual of the shells that I'd seen in Bahía, and without hesitation said that I belonged to the goddess Yemayá. It was impossible that the two seers had reached some kind of accord. This time, Tabra had to swallow her sarcastic comments.

We left the *favela* and on the way back saw a modest café where they sold typical food by its weight. It seemed to me it would be more picturesque to have lunch there rather than eat shrimp cocktail on the hotel terrace, and I asked the driver to stop. He stayed in the van to look after the photographic equipment while the rest of us stood in line at the counter where they were dishing stew onto paper plates with a wooden spoon. I don't know why I walked outside, followed by Lori and Amanda; maybe to ask the driver if he wanted something to eat. When I looked out the door, I noticed that the street, which had been bustling with traffic and activity, had emptied; no cars were driving by, the shops all seemed to be closed, people had disappeared. Across the street, some thirty feet away, a young man wearing blue pants and a short-sleeved T-shirt, was waiting at the bus stop. At his back, a man not unlike him was advancing toward him; he, too, was young, wearing dark pants and a similar T-shirt. He had a large pistol in his hand, making no effort to hide it. He raised the weapon, aimed at the first young man's head, and fired. For an instant I didn't know what had happened; the shot wasn't explosive like those in the movies, but a muffled kind of cough. Blood gushed from the victim's head before he fell. And when he hit the ground, the murderer shot four more times. Then, calm and defiant, he walked on down the street. Like an automaton, I ran toward the man who lay bleeding on the ground. He shuddered convulsively once or twice, and lay quiet, as a pool of luminous blood grew around him. Before I could kneel to help him, my friends and the driver, who had hidden in the van during the crime, dragged me to our vehicle. In minutes the street was again filled with people. I heard screams, horns, and saw clients come running out of the restaurant.

The Brazilian journalist forced us to get into the van and told the driver to take us to the hotel by way of the side streets. I thought she wanted us to avoid the blocked traffic that undoubtedly would follow, but she explained that it was a strategy to avoid the police. It took us forty minutes to get back to the hotel, which seemed eternal. On the way I was assailed by images of the military coup in Chile, the dead in the streets, the blood, the sudden violence, the sensation that at any moment something fatal could happen, that no one was safe anywhere. The press was waiting at the hotel with television cameras; inexplicably, they had learned what had happened, but my editor, who was also there, didn't allow us to speak with anyone. She led us quickly to one of our rooms and ordered us to stay inside until they could take us directly to the airplane. The assassination might have been a settling of accounts among criminals, but the way it happened, outdoors and in broad daylight, it seemed more like one of the infamous executions by agents of the police, which at that time had taken the law into their own hands with full impunity. The press and the public commented on these kinds of events but there had never been proof, and had there been any, it would have opportunely disappeared. When they learned that a group of foreigners—I among them, my books are reasonably well known in Brazil—had witnessed the crime, the journalists supposed that we could identify the assassin. If that were the case, we were told, any number of people would make it their business to see that we didn't testify. Within a few hours we were on the return flight to California. The journalist and the driver had to go into hiding for weeks.

This incident was Lori's test of fire. When we scrambled into the van, Lori was trembling in Amanda's arms. I admit that the sight of a man bleeding to death from five pistol shots is terrible, but Lori had been assaulted two or three times in New York and had traveled through much of the world, this wasn't the first time she had found herself in a violent situation. She was, however, the only one who was having an extreme reaction; the rest of us were mute, but processing it. Her reaction was so intense that when we reached the hotel, we had to call a doctor to give her a tranquilizer. That serene young woman who through all the previous weeks had smiled under pressure and had demonstrated good humor when inconvenienced, had dared to swim in the piranha infested river, had firmly dealt with the drunken Russians who though they had treated Tabra and me with the respect due two Ukrainian grandmothers had pressed Amanda and her with unwelcome attention, had crumbled at the time of those five bullets. Perhaps Lori could assume responsibility for my three grandchildren and hold her own with our strange family without bodily injury, but when I saw her in that state, I realized that she was more fragile than she had appeared at first view. She would need a little help.

Matchmaker

The Amazon fired my imagination. I finished writing *Aphrodite* in a few weeks and added erotic recipes from Dadá's kitchen in Bahía as well as others invented by my mother, and then I asked Lori to design the book, a good way to lower her defenses.

Amanda was my accomplice. Once the three of us went to a Buddhist retreat, at Lori's initiative, and, after long sessions of meditation, had ended up sleeping on pallets in cells with rice paper walls. We had to sit for hours on *safus*, round, hard cushions that are part of the spiritual practice. Whoever can survive the cushion is already halfway to illumination. This torture was interrupted three times a day in order to eat a variety of grains and take slow walks, circling a Japanese garden of dwarf pines and rigorously arranged stones.

In complete silence. In our austere cell we choked back our laughter with the *safus*, but a woman with gray braids and limpid eyes came to remind us of the rules. "What kind of religion is this that doesn't allow you to laugh?" Amanda asked. I was a little disturbed because Lori seemed to enjoy this little corner of peace and murmurs, which possibly would fit well with Nico's even temperament but was not compatible with the task of raising three children. Amanda explained that Lori had lived two years in Japan and still had a Zen remora attached to her, but not to worry, that could be cured.

I invited Lori to have dinner with Amanda and Tabra at our house, and introduced Nico and the other two children, who, compared with Andrea, seemed bland. I had told Lori that Nico was still angry about the divorce, and it would not be easy for him to find a partner, since no woman in her right senses would want a man with three runny-nosed kids. To Nico I commented in passing that I had met an ideal woman for him, but since she was older than he was, and already had a boyfriend of sorts, we would have to keep looking. "I think that's up to me," he replied, smiling, but a shadow of panic flashed across his face. I confessed the plan to Willie—he had already guessed it anyway—and instead of repeating his usual warning for me to keep my nose out of it, he put extra effort into preparing an appealing vegetarian meal for Lori; he had seen her and said that she had class and would fit very well into our clan. You would have liked her, too, daughter; you have a lot in common. During dinner, Lori and Nico did not exchange a single word; they didn't even look at each other. Amanda and Tabra agreed with me that we had failed miserably, but a month later my son confessed that he'd been out with Lori several times. I can't understand how they kept that from me for a whole month.

"Are you two in love?" I asked.

"I think that's a little premature," he replied with his habitual caution.

"Love is never premature, particularly at your age, Nico."

"I'm only thirty!"

"Thirty, you say. But it was only yesterday that you were breaking bones on your skateboard and firing eggs at people with your slingshot! The years fly by, son, there's no time to lose."

Years later, Amanda told me that the day after Nico met Lori, my son planted himself at the door of her office building with a yellow rose in his hand, and when finally she came out to go to lunch and found him standing there like a post in the hot sun, Nico told her that he was just "passing by." He doesn't know how to lie; his blush betrayed him.

Soon the man Lori was having an affair with, a rather famous travel photographer, quietly disappeared beyond the horizon. He was fifteen years older than she, thought he was irresistible to women, and in fact may have been before vanity and the years made him a little pathetic. When he wasn't on one of his excursions to the ends of the earth, Lori moved into his apartment in San Francisco, a garret with no furniture but with a superb view, where they shared a strange honeymoon that seemed more like a pilgrimage to a monastery. With good humor she tolerated the man's pathological need for control, his bachelor manias, and the lamentable fact that the walls were covered with pictures of scantily clad young Asian women whom he'd photographed when he was not in the ice of Antarctica or the sands of the Sahara. Lori had to absorb his rules: silence, bows, remove shoes, touch nothing; no cooking because the smells bothered him, no telephones, and certainly no permission to have visitors; that would have been a major lack of respect. She had to walk on tiptoes. The only plus about this fine fellow was that he was often out of town. What did Lori see in him? Her women friends couldn't understand. Fortunately, she was beginning to tire of competing with the Asian girls, and she was able to leave him with no sense of guilt when Amanda and other friends took on the task of ridiculing him while praising the real and imaginary virtues of Nico. When she bid the photographer farewell, he told her not to show up at any of the places where they'd been together. I remember the moment when Nico and Lori's love was made public. One Saturday he left the children with us—in their minds there was nothing better than sleeping with their grandparents and stuffing themselves with sweets and television—and came to pick them up on Sunday morning. One look at his scarlet ears—the color they get when he wants to hide something from me—and it was clear to me that he had spent the night with Lori and, knowing him, that things were getting serious. Three months later they were living together.

The day that Lori brought her belongings to Nico's house, I left a letter on her pillow, welcoming her to our tribe and telling her that we'd been waiting for her, that we'd known she was out there somewhere, and it had only been a question of finding her. In passing, I gave her a bit of advice that had I put in practice myself would have saved a fortune in therapists: Accept the children the way we accept trees—with gratitude, because they are a blessing—but do not have expectations or desires. You don't expect trees to change, you love them as they are. Why hadn't I done that with my stepsons? If I had accepted them as naturally as I would a tree, I would have had fewer tiffs with Willie. Not only did I try to change them, I assigned myself the thankless role of guardian for the whole family during those years they were dedicated to heroin. I added in that missive that it is futile to try to control children's lives, or to protect them from all harm. If I hadn't been able to protect you from death, Paula, how could I protect Nico and my grandchildren from life? More good advice I don't practice.

In order to live with Nico, Lori had to change her life completely. From being a sophisticated young single woman in a perfect apartment in San Francisco, she had to change into a suburban wife and mother, with all the boring tasks that come with it. She previously had every detail of her life under control, and now she had to keep her head above water in the turbulent disorder of a houseful of children. She got up at dawn to do her household chores, then drove to San Francisco to her design studio or spent hours on the road to meet clients in other cities. She didn't have much time left for reading, her passion for photography, travel, numerous friends, or her yoga and Zen, but she was in love and she assumed the role of wife and mother without a word. The family quickly absorbed her. She didn't know then that she would have to wait ten years—until the children could look after themselves—before with conscious effort she regained her old identity.

She transformed Nico's life and his dwelling. The unrefined furniture, artificial flowers, and discordant paintings all disappeared. She remodeled the house and planted a garden. She painted the living room, which before had looked like a dungeon, Venetian red—I nearly fainted when I saw the sample but it turned out really well—bought light pieces of furniture and tossed silk cushions here and there, the way you see them in magazines of home décor. She hung photos of the family in the bathrooms and added thick towels and candles in tones of green and magenta. In their bedroom she had orchids, necklaces hanging on the walls, a rocker, antique lamps with lace shades, and a Japanese trunk. Her hand could be seen in every detail, even the kitchen, where reheated pizzas and bottles of Coca-Cola were replaced with the Italian recipes of a great-grandmother in Sicily, tofu, and yogurt. Nico likes to cook—his specialty is that Valencia paella you taught him to make—but when he was on his own he hadn't had time or spirit for pots and pans. With Lori beside him, he recaptured his skills. Lori brought a feeling of home that had been greatly needed, and Nico soaked it up; I had never seen him so content and playful. They held hands and kissed behind doors, spied on by the children, while Tabra, Amanda, and I congratulated ourselves on our selection. Occasionally I went by their house at breakfast time, because the spectacle of an apparently happy family comforted me for the rest of the day. The morning light would be flooding into the kitchen; you could see the garden through the window, and a little farther in the distance, the lake and wild ducks. Nico would be cooking stacks of pancakes, Lori cutting fruit, and the children, smiling, disheveled, and still in pajamas, would be wolfing everything down. They were still very young; Nicole had just turned three, and their hearts were open. The atmosphere was festive and tender, a relief after the drama of illnesses, deaths, divorce, and fights we had borne for so long.

Mother-in-Law from Hell

I said that I "sometimes" dropped by, but the truth is that I had a key to Nico and Lori's house and had established bad habits. I went at any hour, without previous notice, interfered in the lives of my grandchildren, treated Nico as if he were still a boy . . . in sum, I was a pernicious mother-in-law. Once I bought a rug, and without asking permission moved all the furniture and put it in their living room as a surprise. I didn't stop to think that if anyone rearranged the décor of my house they'd get a crack on the head in way of thanks. You, Paula, would have given back the rug and dressed me down at the same time—though I wouldn't have dreamed of leaving a nine-by-fifteen Persian rug in your home to begin with. Lori thanked me, pale but courteous. Another time, I bought some nice dishcloths to replace the ones they were using, and threw the old ones in the trash, never suspecting that they had belonged to Lori's dead grandmother, and that she had treasured them for twenty years. Using the excuse of waking my grandchildren with a kiss, I would slip into my son's house at dawn. It wasn't a rare event for Lori, coming nearly naked from her shower, to find her mother-in-law in the hall. And as if that weren't enough, I was secretly visiting Celia, which in all truth was a betrayal of Lori, though I was incapable of seeing it in that light. By one of those jokes of fate, Nico invariably found out. Although I saw Celia and Sally much less, I never lost contact with them, sure that eventually things would smooth out. The lies and omissions on my part, and resentment on Nico's, were building up. Lori was confused; everything around her was in flux, nothing was clear and concise. She didn't understand that my son and I had been absolutely open with each other except in the matter of Celia. It was she who insisted on the truth; she said that she couldn't bear that slippery terrain, and asked how long we were going to avoid a healing confrontation. It's superfluous to say that we had several.

"I have to maintain some kind of relationship with Celia," Nico explained. "I hope it can be civilized, but minimal. She's abrasive, she aggravates me with her bad temper and the fact that she is constantly changing the rules on me."

"I understand that, but I'm not in a similar position. You're my son and I adore you. My friendship with Celia has nothing to do with you or with Lori."

"Yes, it does, Mamá. It makes you feel bad to watch her going through bad times. But can't you think about me? Aren't you forgetting that she provoked this situation? She was the one who split up this family and she has to pay the consequences."

"I don't want to be a half-time grandmother, Nico. I also need to see the children during the weeks they're with Celia and Sally."

"I can't stop you, but I want you to know that I'm hurt and angry, Mamá. You treat Celia like the prodigal son. She will never take Paula's place, if that's what you're trying to do. You feel indebted to her because she was with you when my sister died, but I was there, too. The closer you get to Celia, the farther you will drive Lori and me away. It's inevitable."

"Oh, Nico. There aren't any fixed rules for human relationships, they can be reinvented, we can be original. With time the anger will wane and the wounds heal—"

"Yes, but that won't make me feel any better about Celia, that I can promise. So are you really close to my father, or Willie to his ex-wives? This is a divorce. I want to keep Celia at a prudent distance, so I can relax and live my life."

One memorable night, Nico and Lori came to tell me that I was too much in their lives. They tried to do it with delicacy, but just the same the trauma nearly gave me a heart attack. I threw a childish fit, convinced that they had dealt me a great injustice. My son was throwing me out of his life! He ordered me not to countermand his instructions in regard to the children: no candy before dinner, no money or gifts unless it was a special occasion, no watching television till midnight. So what is a grandmother good for? Did he intend to sentence me to a solitary existence? Willie seemed to be squarely with me, but I think that deep down he was laughing a little. He made me see that Lori was as independent as I am, that she had lived alone for years and was not used to having people troop through her home uninvited. And what on earth made me take a rug to a designer?

As soon as I could control my desperation, I called Chile and talked with my parents. At first they didn't understand very well what the problem was, since in Chilean families relations tend to be what I had imposed on Nico and Lori, but then they remembered that customs are different in the United States. "Child, we come into this world to lose everything. It costs nothing to let go of material things, what's difficult is to give your loved ones their freedom," my mother told me with sorrow. That was her fate; none of her children or grandchildren lived near her. Her words unleashed another torrent of complaints, which Tío Ramón interrupted with the voice of reason to point out that Lori must have made a lot of concessions to be with Nico: moving from the city and her own place, modifying her lifestyle, adapting to three stepchildren and a new family, and on and on, but the worst was the overwhelming presence of her mother-in-law. That couple needed air and space to cultivate their relationship without my hanging over every move they made. He recommended that I make myself scarce, and added that children must be separated from the mother or they will be infantile forever. Whatever my good intentions, he said, I would always be the matriarch, a position that others undoubtedly resent. He was right: my role in the tribe is disproportionate, and I lack the restraint of an Abuela Hilda. Willie's description of me is "a hurricane in a bottle."

Then I remembered a Woody Allen film in which his mother, an overpowering old woman with a pile of dyed red hair and the eyes of an owl, accompanies him to a theater program. The magician says he will make someone disappear and asks for a volunteer from the audience, and without thinking twice, the old lady goes up on the stage and crawls into a trunk. The illusionist performs his trick, and she vanishes forever. They look for her inside the magic trunk, in the wings, inside and outside the building. Nothing. At last the police, detectives, and firemen arrive, but their attempts to find her are futile. Her son, thrilled, believes that he has finally rid himself of her, but the evil old woman appears in the sky, ensconced on a cloud, omnipresent and infallible, like Jehovah. That was me, it seems, like the mother in Jewish jokes. Using the excuse of helping and protecting my son and grandchildren, I had turned into a boa constrictor. "Concentrate on your husband; that poor man must be good and sick of your family," my mother added. Willie? Sick of me and my family? I'd never thought of that. But my mother was right. Willie had put up with your death, Paula, and my long mourning, the problems with Celia, Nico's divorce, my long absences on trips, my obsessive dedication to writing, which kept me always with one foot in another dimension, and who knows what else. It was time for me to let go of the cartload of people I'd been dragging with me since I was nineteen and focus more on my husband. I was shattered; I threw the key to Nico's house into the trash and prepared to remove myself from his life—well, not disappear entirely. That night I cooked one of Willie's favorite dishes, seafood pasta, opened our best bottle of white wine, and waited for him wearing red. "Is something the matter?" he asked when he arrived, dropping his heavy briefcase on the floor.

Lori Comes in Through the Front Door

That was a period of many adjustments in the family's relationships. I think that my need to create and hold together a family, more accurately, a small tribe, had been a part of me since my marriage when I was twenty years old, had grown stronger on leaving Chile— when my first husband and children reached Venezuela we had no friends or relatives except my parents, who had also sought asylum in Caracas—and was consolidated when I found myself an immigrant in the United States. Before I came into Willie's life he had no idea what a family was; he lost his father when he was six, his mother retired into a private spiritual world to which he had no access, his first two marriages failed, and his children had very early set off on the path of drugs. At first, it was difficult for Willie to understand my obsession with gathering my children around me, to live as close to them as possible and to add others to that small base to form the large, united family I had always dreamed of. Willie considered it a romantic fantasy, impossible to carry out on the practical level, but in the years we've lived together not only has he realized that this is the way people live in most parts of the world, but also that he likes it. A tribe has its inconveniences, but also many advantages. I prefer it a thousand times to the American dream of absolute individual freedom, which, though it may help in getting ahead in this world, brings with it alienation and loneliness. For those reasons, and for all we had shared with Celia, losing her was a hard blow. It had wounded us all, and it had completely disrupted the family we had worked so hard to bring together. Even so, I missed that girl.

Nico tried to keep Celia at a distance, not just because that's the normal thing to do in a divorce, but because he felt that she was invading his territory. I hadn't measured the depths of his feelings or thought it was necessary to choose between them; I thought that my friendship with Celia had no connection to him. I didn't give him the unconditional support that as his mother I owed him. He felt that I had betrayed him, and I can imagine how much I must have hurt him. We weren't able to speak openly because I avoided the truth; his eyes would fill with tears and the words wouldn't come out. We loved each other very much but we didn't know how to handle a situation in which inevitably we wounded each other. Then Nico wrote me several letters. Alone with the page he could express himself and I could hear him. How badly we needed you, Paula! You always had the gift of seeing things clearly. At last we decided to go together for therapy, where we could talk and cry and take each other's hand and forgive.

While your brother and I were trying to better understand our relationship, delving into the past, and into the real person each of us was, Lori was focused on curing Nico of the wounds left by his divorce. She made him feel loved and desired, and that transformed him. They took long walks, went to museums, theaters, and good movies; she introduced him to her friends, nearly all artists, and piqued his interest in traveling, something she had done since she was a girl. She gave the children a calm home, as Sally did in the other house. Andrea wrote in a composition at school that "having three mothers is better than just one."

In the matter of a year or two, Lori's office was no longer profitable. Clients believed that they could replace the artist's vision with a disc, and thousands of designers were out of a job. Lori was one of the best.

She had done such a remarkable job with my book *Aphrodite* that my editors in twenty countries used the design and illustrations she had chosen. Because of the design, not the content, the book attracted attention. It wasn't a subject to be taken seriously, and, in addition, a new drug had just been launched that promised to end male impotence. Why study my ridiculous manual and serve oysters in your negligee if all that was needed was a blue pill? The tone of the letters some readers sent me regarding Aphrodite was noticeably different from those I received after *Paula.* A seventy-seven-year-old gentleman invited me to participate in hours of intense pleasure with him and his sex slave, and a young Lebanese man sent me thirty pages extolling the advantages of the harem. All this while in the United States the scandal involving President Bill Clinton and a plump White House intern obscured the successes of his government and would later cost the Democrats an election. A soiled dress and a pair of panties came to have more weight in American politics than the outstanding economic, political, and international accomplishments of one of the most brilliant presidents the country has known. His affair precipitated an investigation worthy of the Inquisition, and cost a trifling fifty-one million dollars to taxpayers. During that period, I was asked to be on a live call-in radio program. Someone asked me what I thought about the matter, and I said that it was the most expensive blow job in history. That phrase would follow me for many years. It was impossible to hide from the kids what was going on because even the most obscene details were shown on TV.

"What is oral sex?" Nicole asked, a term she had heard ad nauseam on television. "Oral? That's when you talk about it," replied Andrea, who had the vast vocabulary of every good reader.

I recall that one magazine decided to feature one of my books with an interview and photographs taken in our house, which Lori was to supervise since I didn't understand what the devil they meant to do. Three days in advance, two artists came to take light readings, color samples, measurements, and Polaroids. For the article, seven people came in two vans loaded with fourteen boxes filled with all sorts of things, from knives to a tea strainer. These invasions happen with some frequency, but I never get used to them. In this case, the team included an artistic director and two chefs, who took over the kitchen to prepare a menu inspired by my book. The dishes were produced with mind-numbing slowness; they placed each lettuce leaf as if it were the feather on a hat, precisely in the angle between the tomato and the asparagus. Willie got so nervous he had to leave, but Lori seemed to comprehend the importance of the damned lettuce. In the meantime, the artistic director replaced the flowers in the garden, which Willie had planted with his own hands, with others more colorful. None of this appeared in the magazine, the photos they used were all close shots: half a clam and a lemon slice. I asked why they had brought the Japanese napkins, the tortoise-shell serving spoons, the Venetian lanterns, but Lori shot me a look that said I should keep quiet. This lasted the entire day, and since we couldn't attack the meal before it was photographed, we put away five bottles of white wine, and three red, on empty stomachs. By the end, even the artistic director was stumbling. Lori, who had drunk nothing but green tea, had to carry the fourteen boxes back to the vans.

Lori kept afloat longer than other designers, but the day came when it was impossible to ignore the red numbers in her account books. At that point, I proposed that she take full charge of the foundation I had established on my return from India, inspired by the infant beneath the acacia, something she had been doing part-time for a while. Every year I earmarked a substantial part of my income for the foundation that evolved from your plan to do good, financed by the sale of my books. During the year that you were sleeping you taught me a lot, daughter. Paralyzed and mute, you still were my teacher, just as you had been all the twenty-eight years of your life. Very few people have the opportunity you gave me; to have quiet time and silence to think and remember. I was able to review my past, to become aware of who I am in essence, once vanity was not a factor, and to decide how I want to be in the years I have left in this world. I appropriated your motto: "You have only what you give," and I discovered, to my surprise, that it is the keystone of my contentment. Lori has your integrity and compassion. She could carry out the proposal "to give till it hurts," as you used to say. We sat at my grandmother's magical table and talked for days, until we had a clear outline of our mission: to help the poorest of women by any means within our reach. The most backward and miserable societies are those in which women are oppressed. If you help one woman, her children do not die of hunger, and if families are bettered, the village benefits, but this very evident truth is ignored in the world of philanthropy, where for every dollar destined to programs for women, twenty go to projects for men.

I told Lori about the woman wrapped in the black plastic bag I'd seen crying on Fifth Avenue, and also about Tabra's most recent experience. She had just returned from Bangladesh, where my foundation maintained schools for girls in remote villages and a small clinic for women. Tabra went there with a friend of hers, a young dental hygienist who wanted to offer her services at the clinic. They filled suitcases with medications, syringes, toothbrushes, and any supplies they could collect from dentist friends. When they reached the village there was already a line of patients at the door, a hot building infested with mosquitoes, where there was little to be seen other than the walls. The first woman had several rotten molars, and had been maddened for months by persistent pain. Tabra acted as assistant while her friend, who had never pulled a tooth, anesthetized the woman's mouth with trembling hands and proceeded to extract the bad molars, trying not to faint in the process. When she finished, the unfortunate woman kissed her hands, grateful and relieved. That day they saw fifteen patients and removed nine molars and a variety of other teeth, while in a tight circle outside, the men of the community observed and commented. The next morning Tabra and her friend arrived early at the improvised clinic and found the first patient of the previous day with a face swollen the size of a watermelon. Her husband was with her, irately shouting that they had ruined his wife, and that the men of the village were meeting to take revenge. Terrified, the hygienist gave the woman anti-biotics and painkillers, praying that there would be no fatal consequences.

"What have I done? She's deformed!" she moaned when the two left.

"That wasn't from your procedure," their interpreter informed them. "Her husband beat her last night because she didn't get home in time to fix his meal."

"That's the life most women live, Lori. They are always the poorest of the poor. They do two-thirds of the work in this world, but they own less than one percent of the assets."

Up to that point, the foundation had distributed money on impulse, or yielding to the pressure of a just cause, but thanks to Lori we established priorities: education, the first step to independence in every sense; protection, because too many women are trapped in fear; and health, without which nothing else helps much. I added birth control, which for me has been essential. Had I not been able to decide something as basic as the number of children I would bring into the world, I couldn't have done any of the things I have. Fortunately, the pill was invented or I would have had a dozen little ones.

Lori threw herself into the work of the foundation, and in the process demonstrated that she had been born for the job. She is idealistic and organized, she notices the least detail, and she is a workaholic—no job is too big for her. She made me see that it was no good to distribute money by tossing it up in front of a fan; you had to evaluate the results and support programs for years, the only way that aid does any good. We also had to concentrate our efforts; we could not put patches on something in a remote place where no one supervised, or bite off more than we could chew; it was better to give more to fewer organizations. Within a year she changed the face of the foundation, and I was able to delegate everything to her. All she asks of me is to sign the checks. She has succeeded so admirably that not only has she multiplied the funds we give out, she has built up the capital as well, and now she manages more money than we had ever imagined. Everything goes to the mission we proposed: carrying out your plan, Paula.

Mongolian Horsemen

In the middle of that year I had a spectacular dream, and I wrote it down to tell my mother; we always did that, even though there's nothing as boring as listening to other people's dreams. That's why psychiatrists cost you so much. Dreams are essential to our lives; they help us understand our reality, and bring into the light the things that are buried in the caverns of our souls. I was standing at the foot of a wind-eroded cliff, on a white sand beach with a dark sea and clear blue sky. Suddenly, at the top of the cliff, I saw two enormous warhorses, with riders. Beasts and men were arrayed like Asian warriors of old—Mongolia, China, Japan—with silk standards, ball fringe, plumes, and heraldic adornment: all the splendid paraphernalia of war gleaming in the sun. After an instant's hesitation at the edge of the precipice, the steeds reared, whinnied, and with the glory of centaurs, leaped into the void, forming in the sky a broad arc of cloth, plumes, and pennants. Their daring took my breath away. It was a ritual act, not suicide, a demonstration of bravery and skill. An instant before they touched ground, the horses bent their necks and landed on one shoulder, curled into a ball, and rolled over, raising a cloud of golden dust. And when the dust and the noise subsided, these stallions struggled to their feet in slow motion—the horsemen had not been unseated!—and galloped off down the beach toward the horizon. Days later, when I still had those images fresh in my memory, trying to make sense of them, I ran into a friend who writes books on dreams. She gave me her interpretation, which was not unlike what the shells of the *jogo de búzios* had said in Brazil: a long and dramatic fall had tested my courage, but I had risen and, like the steeds, had shaken off the dust and run on toward the future. In the dream, the mounts had known how to roll and the horsemen how to sit their horses. According to my friend, past trials had taught me how to fall and now I didn't have to fear because I could always land on my feet. "Remember those horses when you feel yourself weaken," she said.

I remembered two days later, when a theater work based on my book *Paula* was premiered. On the way to the theater, we passed the Folsom Street fair in San Francisco. We had no idea that it was the day for the sadomasochists' carnival: blocks and blocks crowded with people in the most outrageous garb. "Freedom! Freedom to do what I want: fuck!" shouted a good man dressed in a monk's cassock open in the front to display a chastity belt. Tattoos, masks, revolutionary Russian hats, chains, whips, hair shirts of every nature. The women had black- or green-painted mouths and fingernails, stiletto-heeled boots, black plastic garter belts, in short, all the symbols of this picturesque culture. There were several monumentally fat women sweating in leather pants and jackets with swastikas and skull decals. Ladies and gentlemen wore rings or studs through their noses, lips, ears, and nipples. I didn't dare look any lower. A young woman with bared breasts was riding on the hood of a '60s car, her hands tied behind her back; another woman, dressed as a vampire, was lashing her chest and arms with a horsewhip. It wasn't a joke; she was badly bruised and her screams could be heard through the entire area. All this was taking place before the amused eyes of a pair of policemen and various tourists taking pictures. I wanted to intervene, but Willie grabbed my jacket, lifted me off the ground, and dragged me away, feet kicking in the air. A half block farther on we saw a fat-bellied giant carrying a dwarf wearing a leash and dog collar. The dwarf, like his master, wore combat boots and nothing else except a sheath of metal-studded black leather on his whacker, held precariously by a few invisible little ties threaded up the crack of his butt. The little guy barked at us but the giant greeted us very amiably, and offered us some candies in the shape of penises. Willie let go of me and stood gaping at the pair. "If I ever write a novel, that dwarf will be my protagonist," he said, totally out of the blue.

The play *Paula* began with the actors in a circle, holding hands, summoning your spirit, daughter. It was so moving that not even Willie could contain his sobs when at the end they read the letter you had written, "To be opened when I die." A slim girl, ethereal and graceful, played the leading role, dressed in a white shirt. Sometimes she lay on a cot in a coma; other times her spirit danced among the actors. She didn't speak until the end, to ask her mother to help her die. Four actresses represented different moments of my life, from child to grandmother, and passed from hand to hand a red silk shawl that symbolized the narrator. One actor played Ernesto and Willie; another was Tío Ramón, and he drew laughs from the audience when he declared his love for my mother, or explained to Paula how he was a direct descendent of Jesus Christ—just go look up the tomb of Jesús Huidobro in the Catholic cemetery in Santiago. We left the theater in silence, feeling that you were floating among the living. Did you ever imagine, Paula, that you would touch so many people?

The next day we went to the forest of your ashes to greet you and Jennifer. Summer had ended, the ground was carpeted with crunching leaves, some trees had dressed in the colors of fortune, from dark copper to shining gold, and in the air was the promise of the first rain. We sat on a redwood tree trunk in a chapel formed by the high treetops. A couple of squirrels were playing with an acorn at our feet, giving us sideways glances, not at all afraid. I could see you, whole, before your illness wrought its devastation: at three, singing and dancing in Geneva, at fifteen, receiving a diploma, at twenty-six, dressed as a bride. I sat thinking about my dream and the horses that fell and rose again. I have fallen and risen many times in my life, but no fall was as hard as the one of your death.

A Memorable Wedding

Two years after the first night they spent together, in January of 1999, Nico and Lori were married. Up till then she had resisted because she couldn't see why it was necessary; he, however, thought that the children had been through a lot and would feel more secure if he and Lori were husband and wife. The children had seen Celia and Sally always together and didn't question their love, but I think they were afraid that if we were careless Lori would get away. Nico was right; the children were happier about the decision than anyone. "Now Lori will be with us more," said Andrea. They say that it takes eight years to adapt to the role of stepmother, and that the most difficult of all is the task of the childless woman who comes into the life of a man who is a father. It wasn't easy for Lori to change her life and accept the children; she felt invaded. Nevertheless, she took over all the thankless tasks, from washing clothes to buying shoes for Andrea, who wore only green plastic sandals—and not just any sandals, they had to be from Taiwan. Lori killed herself working to be the perfect mother, never overlooking a single detail, but she really didn't need to take such pains, since the children loved her for the same reasons the rest of us did: her laugh, her unconditional affection, her friendly jokes, her tempestuous hair, her boundless goodness, her way of being there in good and bad times.

The wedding was a joyful ceremony in San Francisco, which culminated with a group class in swing, the first time Willie and I had danced together since the humiliating experience with the Scandinavian instructor. Willie, in his dinner jacket, looked like Paul Newman in one of his films, though I don't remember which one. Ernesto and Giulia came from New Jersey, Abuela Hilda and my parents from Chile, but Jason couldn't get away from work. He was still single, though he was not wanting for women to keep him company for one night. According to him, he was looking for someone as reliable as Willie.

We met Lori's friends, who came from the four points of the compass. Over time, several of them became Willie's and my best friends, despite the difference in ages. Later, when we received our photos of the party, I realized that they all looked like magazine models; I have never seen a group of such beautiful people. Most were talented, unpretentious artists: designers, graphic artists, caricaturists, photographers, filmmakers. Willie and I immediately were friends with Lori's parents, who did not see in me any incarnation of Satan, as Celia's parents had, even though in my brief words at the reception I made the bad mistake of alluding to carnal love among our children. Nico still hasn't forgiven me. The Barras, uncomplicated, loving people, are of Italian origin and have lived for more than fifty years in the same house in Brooklyn where they brought up four children a block away from the old mansions of the mafiosi, which can be distinguished from others in the neighborhood by the marble fountains, Greek columns, and statues of angels. Lori's mother, Lucille, is slowly losing her eyesight, but she makes light of it, not so much because of pride as not to be a bother. In her house, which she knows by memory, she moves with assurance, and in her kitchen she is invincible; she continues to prepare by touch the complicated recipes handed down from generation to generation. Her husband, Tom, a storybook grandfather, embraced me with genuine affection.

"I've prayed a long time that Lori and Nico would get married," he confessed.

"So they wouldn't go on living in mortal sin?" I asked as a joke, knowing that he is a practicing Catholic.

"Yes, but more than anything because of the children," he answered, absolutely serious.

Tom had owned a neighborhood pharmacy before he retired. That had prepared him for stress and fright, since he'd been assaulted on several occasions. Although he's no longer young, he still shovels snow in the winter and climbs a folding ladder to paint ceilings in the summertime. He has steadfastly battled the rather peculiar renters who through the years have occupied a small apartment on the first floor of their house, such as the weightlifter who threatened him with a hammer, the paranoid man who stacked newspapers from floor to ceiling and left barely an ant corridor from the door to the bathroom and from there to the bed, or the third renter who exploded—I can't think of another word to describe what happened— and left the walls covered with excrement, blood, and organs, which Tom, of course, had to clean up. No one could explain what had happened because no trace of explosives was found; my theory is that it must have been something like the phenomenon of spontaneous combustion. Despite these and other macabre experiences, Lucille and Tom have maintained their trust in humanity.

Sabrina, who was already five years old, danced the entire night, clinging to various people, while her vegetarian mothers took advantage of the occasion to try, surreptitiously, the lamb and pork chops. Alejandro, in a grave-digger suit and tie, presented the rings, accompanied by Andrea and Nicole, dressed like princesses in amber satin that contrasted with the bride's long purple gown. Lori was radiant. Nico was very full of himself in black and a Mao shirt, with his hair tied back at the nape of his neck and looking more than ever a sixteenth-century Florentine nobleman. It was an ending I will never be able to use for my novels: they got married and lived happily ever after. That's what I told Willie as he danced to the swing band and I tried to follow. The man leads, as that Scandinavian always said.

"I can die right here of a timely heart attack because my labor in this world is now complete: I have placed my son," I told him.

"Don't even think of it; it's now that they're going to need you," he replied.

Toward the end of the evening, when the guests were beginning to leave, I crawled under a long cloth-covered table with a dozen children drunk on sugar and revved up by the music, their clothes in tatters from all the running around. The word had spread among them that I knew all the stories there were; all they had to do was ask for one. Sabrina wanted the story to be about a mermaid. I told them about the tiny siren who fell into a whisky glass and was drunk by Willie without his realizing. The description of the voyage of the unfortunate little siren through their grandfather's organs, the vicissitudes of swimming through his digestive system, where she encountered every manner of obstacle and repugnant hazard, until she was floating in his urine and emptied into a sewer and from there into San Francisco Bay, left them speechless. The next day, Nicole, wild-eyed, came to tell me she hadn't liked the story of the little siren at all.

"Is it a true story?" she asked.

"Not everything is true, but then not everything is false either."

"How much is false and how much is true?"

"I don't know, Nicole. The essence is true, and in my work as a storyteller that's all that matters."

"There aren't any mermaids, so everything in your story is a lie."

"And how do you know that the siren wasn't a bacterium, for example."

"A mermaid is a mermaid, and a bacterium is a bacterium," she replied, indignant.

To China in Search of Love

Tong accepted a social invitation for the first time in the thirty years he'd worked as bookkeeper in Willie's office. We had resigned ourselves to not inviting him, since he never came, but Nico and Lori's wedding was an important event, even for a man as introverted as he. "Is obligation to go?" he asked. Lori said yes it was, something no one had dared tell him before. He came alone; finally his wife had asked for the divorce, after years and years of sleeping in the same bed without speaking. I thought that in view of the success I'd had with Nico and Lori, I could also look for a girlfriend for Tong, but he informed me that he wanted a Chinese woman, and in that community I was sadly lacking in contacts. In Tong's favor was the fact that San Francisco's Chinatown is the most heavily populated and famous Chinese enclave in the Western world, but when I suggested looking there, he explained that he wanted a woman who had not been contaminated by America. He was dreaming of a submissive wife, eyes always cast to the ground, who would cook his favorite dishes, cut his fingernails, give him a son, and in passing serve her mother-in-law like a slave. I don't know who had put that fantasy in his head, probably his mother, that tiny old lady we all feared. "Do you believe there is such a woman left in this world, Tong?" I asked, perplexed. For an answer, he led me to the computer screen and showed me an unending list of photos and descriptions of girls ready to marry a stranger in order to flee their country or their family. They were classified by race, nationality, and religion, and should the inquirer be more demanding, even the size of their bras. If I had known that this supermarket of female bargains existed, I wouldn't have agonized so over Nico. Although, thinking it over, it's better I didn't. I would never have found Lori on those lists.

The search for Tong's future bride turned out to be a long and complicated office project. At that time we had divided the old Sausalito whorehouse into Willie's law office, my office on the first floor, and Lori's on the second, where she managed the foundation. Lori's elegant touch had changed that house as well, which now was resplendent with framed posters of my books, Tibetan rugs, blue and white porcelain jardinières filled with plants, and a complete kitchen where there was always everything we needed to serve tea, as if we were at the Savoy. Tong gave himself the task of selecting the candidates, which we then criticized: this one has mean eyes, this one is evangelical, this one paints her face like a whore, and so on. We didn't allow Tong to be impressed by appearance; photographs do lie, as he knew very well seeing that Lori had computer-enhanced his portrait; she had made him taller, younger, and with whiter skin, which seems to be appreciated in China. Tong's mother installed herself in the kitchen to compare astral signs, and when finally a young Cantonese nurse emerged who seemed ideal to all of us, she went to consult an astrologer in Chinatown, who gave his approval as well. A round face smiled from the photograph, red cheeks and bright eyes that made you want to kiss her.

After a formal correspondence between Tong and the hypothetical bride, which lasted several months, Willie announced that they would go to China to meet her. I couldn't go with them, though I was dying of curiosity. I asked Tabra to stay with me because I don't like to sleep alone. My friend's business was in the black again. She wasn't living with us anymore; she had found a house, small, but with a patio that looked out toward golden hills, where she could create the illusion of the isolation that was so important to her. I'm sure that living with our tribe must have been torture for her. She needs solitude but she agreed to stay with me while Willie was gone. For a while she stopped going out on blind dates because she was working day and night to get out of debt, but she never stopped hoping for the return of her Plumed Lizard, who did appear on the horizon from time to time. Suddenly his recorded voice on the answering machine would order, "It's four thirty in the afternoon; call me before five or you will never see me again." Tabra would get home at midnight, bone-weary, and find this charming message and be upset for weeks. Fortunately, her work forced her to travel, and she had interludes in Bali, India, and other distant places from which she sent me delicious letters filled with adventures and written with that fluid sarcasm that is her trademark.

"Sit yourself down and write a travel book, Tabra," I begged her more than once.

"I'm an artist, I'm not an author," she said defensively. "But if you can make necklaces, I suppose I can write a book."

Willie took his heavy suitcase of cameras to China and returned with some very good photographs,

especially portraits of people, which is what most interests him. As always, the most memorable photo is the
one he didn't get to take. In a remote village in Mongolia, where he had gone by himself because he wanted
to give Tong the opportunity to spend a few days with
the proposed bride without him as a witness, he saw
a hundred-year-old woman with bound feet, feet like
girls had once suffered in that part of the world. He
went up to her and tried to ask with signs if he could
take a photo of her diminutive "golden lilies," but the
centenarian ran away as fast as her tiny deformed feet
would take her, screaming. She had never seen anyone
with blue eyes and thought that Death had come to
carry her away.

The trip was nonetheless a success, according to
my husband, because Tong's future bride was perfect,
exactly what his bookkeeper was looking for: timid,
docile, and unaware of the rights women enjoy in
America. She seemed healthy and strong, and surely
she would give him the desired male child. Her name
was Lili and she earned her living as a surgical nurse,
sixteen hours a day, six days a week, for a salary equivalent to two hundred dollars a month. "No wonder she
wants to get out of here," Willie commented, as if living
with Tong and his mother would be any easier.

Stormy Weather

I got ready to enjoy a few weeks of solitude, which I planned to use on the book I was writing about California in the days of the gold rush. I had been putting it off for four years, though I already had a title, *Daughter of Fortune*, a mountain of historical research, and even the image for the jacket. The protagonist of the novel is a young Chilean girl, Eliza Sommers, born around 1833, who decides to follow her lover, who has left to join in the chase for gold. For a young girl of the time, a journey of that magnitude was unthinkable, but I believe that women are capable of amazing exploits for love. Eliza would never have thought of crossing half the world for the lure of gold, but she never hesitated to do it in order to find the man she'd lost. However, my plan to write it in peace did not work out because Nico wasn't well. To have a couple of wisdom teeth pulled, it had been necessary to give him a general anesthesia for a few minutes, something that tends to be dangerous for people with porphyria. He got out of the dentist's chair, walked to the reception area where Lori was waiting, and felt the world going black. His knees buckled and he fell backward, stiff as a board, striking his neck and back against the wall. He lay on the floor unconscious. It was the beginning of months of suffering for him and anguish for the rest of the family, especially for Lori, who didn't know what was happening to him, and for me, who knew all too well.

My most tragic memories swirled up in furious waves. I had thought that after going through the experience of losing you, nothing could ever move me that much again, but just the hint that possibly something similar was happening to my son, my remaining child, rocked me off my feet. I had a weight in my chest like a rock, crushing me, and making me short of breath. I felt vulnerable, raw, on the verge of tears every moment. At night, while everyone slept, I heard sounds in the walls of the house, long moans from the doorways, sighs in the unoccupied rooms. It was, I suppose, my own fear. All the sorrow that had accumulated during that long year of your dying was stalking about in the house. I have a scene forever engraved in my memory. One day I went into your room and saw your brother, his back to the door, changing your diaper as naturally and calmly as he did his children's. He was talking to you, just as if you could understand, of the times in Venezuela when the two of you were teenagers and you would cover for his pranks and save his skin if he got in a jam. Nico didn't see me; I left and softly closed the door. This son of mine has always been with me, we have shared primordial pain, dazzling failures, ephemeral successes; we left everything behind and have begun again in a new place; we have fought and we have helped each other; in other words, I believe we cannot be separated.

Weeks before the accident at the dentist's, Nico had had his annual porphyria tests and the results had not been good: his levels had doubled over the previous year. After his fall, they kept rising at an alarming rate, and the doctor, Cheri Forrester, who never took her eye off him, was worried. Added to the constant pain from the injury to his shoulder, which prevented him from lifting his arms or bending over, were the pressure of work, his relationship with Celia—which was going through a wretched stage—the ups and downs with me—I was frequently failing in my intention to leave him in peace—and an exhaustion so profound that he would fall asleep standing up. He was even speaking in a murmur, as if the effort of breathing was too much. Often crises of porphyria are accompanied by mental disturbances that alter the patient's personality. Nico, who in normal times prides himself on having the happy calm that characterizes the Dalai Lama, was often boiling with anger, but he could hide it thanks to his unusual self-control. He refused to talk about his condition, and he did not want any special consideration. Lori and I limited ourselves to watching him without asking questions, trying not to make him more annoyed than he already was, though we did suggest that at least he resign from his job; it was very far away and it didn't offer him any satisfaction or challenges. We thought that with his calm temperament, his intuition, and his mathematical ability, he could work as a day trader, but that seemed very risky to him. I told him my dream about the horses, and he told me that was very interesting, but that he wasn't the one who had dreamed it.

There was nothing Lori could do in regard to his health problems, but she stood by him and gave him moral support, never weakening, though she was suffering herself. She wanted children, and to do that she had to subject herself to the torment of fertility treatments. When she started living with Nico they had of course talked about children. She couldn't give up the idea of being a mother, she had already put it off too long waiting for a true love. From the beginning, however, Nico had said he didn't want children; in addition to possibly transmitting his porphyria, he already had three. Nico had become a father at a very young age; he hadn't experienced the freedom and adventure that filled Lori's first thirty-five years, and he intended to cherish the love that had fallen into his life, be a companion, a lover, a friend, and a husband. During the weeks the children stayed with Celia and Sally, Nico and Lori were like sweethearts, but the rest of the time they could only be parents.

Lori said that Nico couldn't understand her terrible emptiness; it seemed to her—perhaps with good reason—that no one was ready to remove a piece of the family puzzle to make room for her; she felt like a stranger. She perceived something negative in the air when she mentioned the subject of another child, and I was responsible for a lot of that; it took me more than a year to realize how important being a mother was for her. I tried not to interfere, not wanting to hurt her, but my silence was eloquent: I thought that a baby would rob Nico and her of the little freedom and intimacy they had. I was also afraid it might displace my grandchildren. The last straw for Lori came on Mother's Day; one of the girls made an affectionate card, gave it to her, and then a little later asked her to give it back so she could give it to Celia. To Lori, that was a dagger in the heart, even though Nico explained over and over that the child was too young to realize what she'd done. Lori's sense of duty seemed almost a penalty; she looked after the children and served them with a kind of desperation, as if she wanted to compensate for the fact she couldn't accept them as her own. And they weren't, they had a mother, but if they'd adopted Sally they were equally eager to love her.

During this period several of Lori's friends got pregnant; she was surrounded by half a dozen women priding themselves on their bellies. They spoke of nothing else; the very air smelled of infants, and the pressure grew for Lori because her chances of being a mother were decreasing month by month, something her specialist made clear. Lori was never jealous of her friends; just the opposite; she spent a lot of time and energy taking their portraits, and she put together a collection of extraordinary images on the subject of pregnancy that I hope one day will become a book.

Nico and Lori were going to a therapist, where I suppose they discussed this subject to the point of nausea, but on an impulse Nico called Tío Ramón in Chile, whose judgment he trusted blindly. "How do you expect Lori to be mother to your children if you don't want to be father to hers?" was his answer. It was an argument of pristine fairness, and Nico not only yielded, he was enthusiastic about the idea; the weight of that decision, nonetheless, fell on Lori. She submitted, silent and alone, to the fertility treatments that took their toll on her body and spirit. She, who had always been so careful to eat well, to exercise, and live a healthful life, felt poisoned by that bombardment of drugs and hormones. Her attempts failed again and again. "If science doesn't do it, you have to put it in God's hands," my loyal friend Pía contributed from Chile. But neither her prayers, nor those of my Sisters of Disorder, nor the supplications to you, Paula, produced results, and so an entire year went by.

A New House for the Spirits

At the top of the same hill where our house was located, a piece of land went on sale: two acres, with more than a hundred old oaks and a superb view of the bay. Willie wouldn't leave me in peace until I agreed to buy it, even though to me it seemed an extravagant whim. He put himself in charge of the project, and decided to build the true house of the spirits. "You have the mentality of a *castellana*, Isabel, the mistress of a castle; you need style. And I need a garden," he said. In my mind, moving would be a harebrained idea because the house where we had lived for more than ten years had its history and a much-loved ghost, and I couldn't allow someone we didn't know to live among its walls. Willie, however, turned a deaf ear to my arguments and went right ahead with his project. Every day he climbed the hill to photograph each stage of the construction. Not a single nail was driven without being recorded by his camera, while I, clinging to my old dwelling place, did not want to know anything about the new one. I went with him a few times, as I felt was my duty, but I couldn't read the plans; to me it looked like a tangle of beams and pillars, lugubrious and too large. I asked for more windows and skylights. Willie said I was in love with the old Irishman who put in the skylights, because between the two houses I had ordered nearly a dozen. One more and the roof would have crumbled like crackers. Who was going to clean that ocean liner? It needed an admiral to understand that tangle of tubes and cables, the boilers, the ventilators, and other mechanisms to alter the climate. There were too many rooms and our furniture would float around in that immensity. Willie ignored my evil-hearted objections, but he did pay attention to my suggestion concerning the size of the windows and skylights, and when finally the house was done and all that was left was to choose the color of the paint, he took me to see it.

It was a thrilling surprise. It was much more than a place to live, it was a proof of love, my own Taj Mahal. My lover had imagined a Chilean country estate with thick walls and tile roof, colonial arches, wrought-iron balconies, a Spanish fountain, and a cabaña at the back of the garden where I would write. My grandparents' big house in Santiago, the one that inspired my first book, was never as I described it, not as grand or as beautiful or as luminous as it was in the novel. The house that Willie built is the house I had imagined. It rose proudly on the top of the hill, surrounded by oaks, and in the tiled entry patio stood three palm trees— three tall, slim ladies wearing hats of green plumes— that had been transported with a crane and set into the holes prepared for them. A wooden sign hanging from the balcony read THE HOUSE OF THE SPIRITS. My earlier resistance disappeared in a sigh; I threw my arms around Willie's neck, grateful, and took over. I decided to paint the outside a shade of peach, and inside the color of vanilla ice cream. It looked like a cake, but we hired a woman who was seven months pregnant and outfitted with a ladder, a hammer, a blowtorch, and acid, who attacked the walls, the doors, and the iron, and in the space of a week's time had "aged" them at least a century. If we hadn't stopped her, the house would have been reduced to a pile of rubble before she gave birth in our patio. The result is a historic oddity: a nineteenth-century Chilean house on a twenty-first century California hilltop.

In contrast to me—I always had my suitcase packed to escape—the one time that Willie really was tempted to divorce me was during the move. It's true that I acted like a Nazi colonel, but in two days' time it looked as if we'd been there a year. The entire tribe participated, from Nico with his tool belt for installing lamps and hanging paintings, to friends and grandchildren who put cups and saucers in the cabinets, unpacked boxes, and carried off bags of trash. In that uproar we nearly lost you, Paula. Two nights later, we considered the task finished, and the fourteen persons whose backs we'd broken in the move dined at the "table of the *castellana*," as Willie had called it from the beginning, complete with candles and flowers: shrimp salad, a Chilean stew, and flan. None of that Chinese food ordered by telephone. And so in style we inaugurated a way of life we'd never had till then.

If I was going to delight in my new situation as *castellana*, that was nothing compared to Willie, who needs a view, space, and high ceilings to expand in, a large kitchen for his experiments, a grill for the unfortunate steers he often barbecues, and a noble garden for his plants. Despite the million allergies that have tormented him from childhood, he goes outside several times a day to count the buds on each bush and breathe in great mouthfuls of the fresh aroma of laurel, the sweet perfume of the mint, and the penetrating pungency of pine and rosemary, while the crows, black and wise, mock him from the skies. He planted seventeen virginal rosebushes to replace those he left at the other house. When I first met him, he had seventeen roses in wheelbarrows, transported for years along the roads of divorce and moving, but he set them in terra firma when he surrendered to his love for me. From the first year, he has cut flowers for my *cuchitril*, the only place he can put them, since they're death to him. My friend Pía came from Chile to give her blessing to the house, and brought with her, hidden in her suitcase, a cutting from "Paula's rosebush" that grows beside the little chapel in her garden, and which two years later would delight us with a profusion of pink blossoms. From her town of Santa Fe, where she lives, Carmen Balcells every week sends a prodigious bouquet of flowers, and those, too, I can't put anywhere near Willie. My agent is as lavish with her gifts as the hidalgos of imperial Spain. Once she gave me a suitcase of magical chocolates that two years later still turn up in my shoes or in some handbag; they reproduce mysteriously in the dark.

From May to September we heat the pool to the temperature of soup, and the house fills with our own and other children that materialize out of thin air, and visitors who come without announcing themselves, like the mailman. More than a family, we're a village. Mountains of wet towels, flip-flops, plastic toys; piles of fruit, crackers, cheeses, and salads on the long kitchen counter; smoke and grease on the grills where Willie makes fillets, ribs, hamburgers, and sausage dance. Abundance and uproar, which compensate for the winter months of withdrawal, solitude, and silence, the sacred time for writing. The summer belongs to the women; we sit around in the garden amid the carnival of flowers and bees in their striped yellow suits to tan our legs and watch the children, in the kitchen to try new recipes, in the living room to paint our toenails, and in special sessions to exchange clothes with friends. Nearly all my clothing comes from Lea, an imaginative designer who makes everything for me on the bias, and long; that way it stretches, shrinks, adapts, and is equally good for a battalion of women of different sizes—including Lori, with her model's body, who by now had given up the obligatory all-black uniform of New York and adopted California colors. Even Andrea likes to put on my dresses, but not Nicole, who has a pitiless eye for teenager fashion. Half the family birthdays, and those of many close friends, fall within those summer months, and we all celebrate together. It's the season for partying, gossip, and laughter. The children bake cookies and fix snacks of quesadillas and fruit smoothies. I suppose that in every commune there's one person who takes on the most unpleasant chores; in ours it's Lori. We have to arm-wrestle Lori to keep her from taking on the job of washing the hills of bowls and plates single-handed. If we're not careful, she's capable of mopping the floor on her knees.

The best part was that a month after we moved in, we began to hear the same inexplicable sounds that had waked us in the other house, and when my mother came from Chile to visit, she verified that the furniture moved at night. That was what the house needed to justify its name. We hadn't lost you in the move, daughter.

The moment had come to call Ernesto and Giulia, who for months had been considering the possibility of moving to California to become part of the tribe and live in the house we had vacated and was waiting for them. They had married a couple of years before in a ceremony attended by the families of the bride and groom, including Jason, who still hadn't learned about the brief amorous interlude between Ernesto and Sally. Ernesto, with great embarrassment, would confess that to him much later. Giulia, on the other hand, did know, but she is not the kind of woman to be jealous of the past. The bride, splendid in her simple white satin gown, showed no sign of being aware of the untimely reaction of some of the guests, who nearly ruined the wedding. Ernesto's parents, though enchanted with Giulia, took turns locking themselves in the bathroom to cry because they remembered you. That wasn't true in my case; in fact I was very happy. I have always known that you looked for Giulia yourself so your husband wouldn't be left alone, something you had always joked you would do. Why did you even talk about death, daughter? Did you have premonitions? Ernesto says that you both felt that your love was not meant to last long, that you had to cherish it, enjoy it quickly before it was taken from you.

Ernesto and Giulia's life in New Jersey was comfortable and they both had good jobs, but they felt isolated and so yielded to my invitation to move into our old house. In order to accept that gift, Ernesto needed a job in California, and as he has a guardian angel, he was hired by a company ten minutes away from his new dwelling. It took them a couple of months to sell their apartment and bring their things across the continent in a truck, and they moved in on the same day in May that several years before we had brought you, Paula, from Spain to spend with us the time you had left to live. That timing seemed to me to be a good omen. We were aware of it because Giulia gave me an album in which she had arranged in chronological order the letters I had written you in 1991, when you were a new-lywed in Madrid, and those I sent Ernesto in 1992 when you were ill here in California and he was working in New Jersey. "We will be very happy here," said Giulia when she walked into the house, and I had no doubt they would be.

Strokes of the Pen

We hadn't yet recovered from our brief brush with movie-land fame when *Of Love and Shadows* was premiered, the film based on my second novel. The actress Jennifer Connelly was so like you—slim, fragile, thick eyebrows, long hair—that I couldn't stay to see the end of the film. There is a moment when she is in a hospital bed, and her companion, Antonio Banderas, picks her up in his arms and supports her on the toilet. I remember clearly the identical scene between you and Ernesto, a little before you fell into the coma. The first time I saw Jennifer Connelly was in a restaurant in San Francisco, where we'd agreed to meet. When I saw her coming toward me in her faded jeans, her white starched blouse and ponytail, I thought I was dreaming: she was you, daughter, revived in all your beauty. *Of Love and*

Shadows was filmed in Argentina because they didn't dare make it in Chile, where the legacy of the recent dictatorship still had a lot of weight. I thought it was an honest film and regretted that it came out with so little promotion, although it did circulate later on video and television.

The book is a political adventure based on actual events; it tells of fifteen campesinos who disappeared after being arrested by the military, but it is essentially a love story. On Willie's fiftieth birthday, a friend gave him that book, which he read on his vacation. Later he thanked her for the book with a note that said, "The author understands love the way I do." And that is why, because of the love he perceived in those pages, he decided to come meet me when I passed through northern California on a book tour. At our first meeting he asked me about the protagonists; he wanted to know if they existed or whether I had imagined them, whether perhaps their love survived the duress of exile, and whether they had ever returned to Chile. I hear that question frequently; it isn't only children who want to know how much truth there is in fiction. I began to answer his questions, but after only a few sentences he interrupted me. "No, no, don't say any more. I don't want to know. The important thing is that you wrote it and therefore you believe in that kind of love." Then he confessed that he had always been sure that love like that was possible and that one day he would live it, although up to that time he had not experienced anything even remotely similar. My second novel brought me luck. Because of it I met Willie.

By then, *Daughter of Fortune* had been published in Europe. According to some critics it was an allegory of feminism: Eliza shed her Victorian corset to dive, with absolutely no preparation, into a man's world in which she had to dress as a male in order to survive; in the process she acquired something very valuable: freedom. I didn't have that in mind when I wrote the book; I thought the theme was simply the fever for gold, the tumult of adventurers, bandits, preachers, and prostitutes who gave birth to San Francisco, but that analysis seems valid to me because it reflects my feminist convictions and the desire for freedom that has determined the course of my life. To write the novel I had traveled with Willie across California, soaking myself in its history and trying to imagine what those years were like, the period in the nineteenth century when gold gleamed in the riverbeds and clefts in the rocks, inflaming men's greed. The distances are still great despite the expressways; on horseback or on foot along narrow mountain paths they must have been infinite. The magnificent geography, the forests, the snow-covered peaks, the rivers with racing waters, invite silence and remind me of the enchanted landscape of Chile. The history of my two homelands, Chile and California, and the peoples who inhabit them, are very different, but the scenery and the climate are very much alike. Often when I return home after a trip, I have the impression of having traveled in a circle for thirty years and ended up again in Chile; there are the same winters of rain and wind, the dry, hot summers, the same trees, the cliff-lined coasts, the cold, dark sea, the endless hills, the clear skies.

After *Daughter of Fortune* followed *Portrait in Sepia,* the novel I was writing at that time, which also connects Chile with California. Its theme is memory. I am an eternal transplant, as the poet Pablo Neruda used to say; my roots would have dried up by now had they not been nourished by the rich magma of the past, which in my case had an inevitable component of imagination. Perhaps it isn't only in my case. It is said that the processes of remembering and of imagining are nearly identical in our brains. The plot of the novel is based on something that happened in a distant branch of my family, when the husband of one of the daughters fell in love with his sister-in-law. In Chile this kind of family history is not aired. Even though everyone knows the truth, a conspiracy of silence is woven around it to keep up appearances. That may be why no one wants to have a writer in the family. The setting was a beautiful country estate at the foot of the Andean cordillera, and the protagonists were the finest people in the world, who didn't deserve such suffering. I believe that the pain would have been more bearable if it had been talked about instead of hidden behind a veil, if instead of locking in the secret they had opened doors and windows to let the breeze carry off the bad smell. It was one of those dramas of love and betrayal buried beneath layers and layers of social and religious conventions, as told in a Russian novel. Just as Willie says, behind closed doors there are many family secrets.

I didn't plan the novel as a second part of *Daughter of Fortune*, although they coincide historically, but several characters, such as Eliza Sommers, the Chinese physician Tao Chi'en, the matriarch Paulina del Valle, and others, made their way into the pages of the new book and there was nothing I could do to stop them. When I was about halfway along in the writing, I realized that I could connect those two novels with *The House of the Spirits* and make a kind of trilogy that begins with *Daughter of Fortune* and uses *Portrait in Sepia* as a bridge. The unfortunate thing is that in one of the books Severo del Valle loses a leg in the war, and in the following book he has two; that is, somewhere there is an amputated leg floating in the dense fog of literary errors.

The research that had to do with California was easy, I had done that for the previous book, but the rest had to be done in Chile, with the help of Tío Ramón, who for months dug through history books, documents, and old newspapers. It was a good excuse to visit my parents more often; they had entered the decade of their eighties and were beginning to look more frail. For the first time I thought of the terrifying possibility that one not too distant day I might be left an orphan. What would I do without them? Without the routine of writing my mother? That year, contemplating the proximity of death, she returned the packs of my letters bundled in Christmas wrapping paper. "Here, you keep them, because if I suffer a sudden stroke it wouldn't do for them to fall into someone else's hands," she told me. Ever since then she has given me my letters every year, with my promise in return that when I die, Nico and Lori will burn them in a purifying bonfire. The flames will carry off our sins of indiscretion. In them we spill out anything that crosses our minds and sling mud on everyone. Thanks to my mother's epistolary talent and my obligation to answer her, I have in my hands a voluminous correspondence in which events are kept fresh. That is how I have been able to write this memoir. The purpose of that methodical correspondence is to keep pulsing the cord that has joined us since the instant of my conception, but it is also an exercise to strengthen memory, that ephemeral mist in which recollections dissipate, change, and blend together; at the end of our days it turns out that we have lived only what we can evoke. What I don't write I forget; it is as if it never happened. That's why nothing significant is left out in those letters. Sometimes my mother calls me to tell me something that has affected her in some major way, and the first thing I think to say is, "Write me about it, so it won't fade away." If she dies before I do, which is probable, I will be able to read two letters every day, one of hers and one of mine, until I am one hundred and five years old, and as by then I will likely be deep in the confusion of senility, it will all seem new to me. Thanks to our correspondence, I will live twice.

Labyrinth of Sorrows

Nico recovered from the injury to his back, the porphyria levels began to go down, and he was seriously considering the possibility of getting a different job. In addition, he began to do yoga and practice sports: lift weights he didn't need to, swim to Alcatraz and back in the icy waters of San Francisco Bay, pedal a bike sixty miles uphill, run from one town to another like a fugitive. He developed muscles where there are no muscles, and he could make pancakes in the yoga tree position: standing on one foot, the other placed against the inside of the opposite thigh, one arm raised, and the other mixing, all the while reciting the sacred word *oooom*. One day he came to eat breakfast at our house and I didn't recognize him. The Renaissance prince had been transformed into a gladiator.

All Lori's attempts to have a baby failed, and with great sadness she said good-bye to that dream. She was worn down by the fertility drugs and all the poking around in her body, but that was nothing compared with the pain in her soul. The relationship between Celia and Nico was close to being hostile, which created tension and greatly affected Lori; she felt attacked. She couldn't overlook how rudely Celia treated her, no matter how much Nico repeated his mantra to her: It isn't personal, everyone is responsible for his or her own feelings, and life isn't fair. I don't think that was much help. Fortunately, the two couples kept the children on the margin of their problems whenever possible.

The role of stepmother is a thankless one, and I myself contributed my drop of bile to the legend. There isn't a single good stepmother in either oral tradition or universal literature—except for Pablo Neruda's, whom the poet called *mamadre*. In general there's no appreciation for stepmothers, but Lori put so much into the task that my grandchildren, with that infallible instinct children have, not only love her as much as they do Sally, she is the first person they go to if they need something because she never fails them. Today they can't imagine their lives without those two women. For years they wanted all four parents—Nico, Lori, Celia, and Sally—to live together, if possible in their grandparents' house, but that fantasy has evaporated by now. My grandchildren have spent their childhood going from one family to the other, like three back-packers. When they were with one couple, they missed the other. My mother was afraid that the system would produce an incurable Gypsy disorder, but the kids have turned out to be more stable than most people I know.

That year of 2000 culminated with a simple ritual for saying good-bye to the child that Lori and Nico never had, as well as to other sorrows. One windy afternoon we drove to the mountains, led by one of Lori's friends, a young woman who is the incarnation of Gaia, the earth goddess. We were equipped with flashlights and ponchos, should night surprise us. From high on a hill, Gaia pointed out a pass and, down below in a valley, a large circular labyrinth outlined with stones, perfect in its geometry. We went down a narrow path between gray hills under a white sky crisscrossed by black birds. Our guide told us that we had met to unburden ourselves of certain sorrows, that we had come to ac-company Lori, but that there was no one who didn't have a pain of his or her own to leave there. Nico had brought a photo of you; Willie, one of Jennifer; Lori, a box and a photo of her young niece. We started walk-ing, following paths traced by stones, slowly, each of us at our own pace, while the large funereal birds circled

and cawed in that pale sky. At times we met one another in the labyrinth and I could see that we were all
deeply moved.

In the center was a pile of rocks like an altar, where
other pilgrims had left memories since soaked by the
rain: messages, a feather, wilted flowers, a medallion.
We sat around that altar and deposited our treasures.
Lori placed on it the picture of her niece, who looked
like the child she had wanted so much, with the colors
and scents of her family. She told us that from the time
she was very young, she had planned with her sister to
live in the same neighborhood and raise their children
together; hers would be a little girl, Uma, and a boy
named Pablo. She added that it was her good fortune that Nico was sharing his children with her, and
that she would try to be a loyal friend to them. From
the box she took three flower bulbs and planted them
in the earth. Beside one of them she set a stone for
Alejandro, who likes minerals; by another, a pink glass
heart for Andrea, who had not as yet emerged from her
horrible pink stage; and by the last one, a live worm for
Nicole, who loves animals. Willie silently placed the
photograph of Jennifer on the altar and weighed it with
little stones so the wind wouldn't carry it away. Nico
explained that he was leaving your photograph so you
would accompany their unborn child and all the other

sorrows collected there, but that he did not want to let go of his. "I miss my sister and I always will, for the rest of my life," he said.

After all these years, the sadness of having lost you is still intact, Paula. Only scratch the surface lightly and it bursts forth anew, as fresh as the first day.

A ritual in a labyrinth among the mountains is not enough, however, to supersede a desire to be a mother, no matter how much therapy and will are employed. It is a cruel irony that while other women avoid having children, or abort them, it was Lori's fate to be denied them. She had to resign herself to not carrying a child; even the fantastic process of implanting another woman's fertilized egg in her womb had been for naught, but she was left the recourse of adopting. There are many, many little ones who have no family waiting for someone to provide them a generous home. Nico was sure that adoption would exacerbate Lori's problems: lack of time, too much work, and no privacy. "If she feels trapped now, it would be worse with a baby," he said to me. I couldn't offer any advice. The crossroads at which Lori and Nico found themselves was cursed. Whichever of the two yielded, that one would always feel resentful: she because Nico had denied her something essential, and he because she had forced an adopted child on him.

Nico and I sometimes went out to a café to have breakfast together, to catch up on everyday events and secrets of the soul. For a year, the predominant theme of those private conversations was Lori's distress and the question of adoption. Nico could not understand that motherhood was more important than the love between them, which was being threatened by her obsession. He told me that they had been born to love each other, that they complemented each other in every way, and that they had the resources to live an ideal life, but instead of appreciating what they had, Lori was suffering for what they didn't have. I explained to Nico that without that need that engulfs us women our species wouldn't have survived. There is really no sensible reason to subject your body to the prodigious effort of carrying and giving birth to a child, to defend it like a lioness at the cost of your own life, to devote every instant to it for years and years until it can get along on its own, and then, having lost it, watch from afar with nostalgia . . . all children do break away sooner or later. Nico argued that the business of being a mother is neither that absolute nor that clear; some women lack that biological imperative

"Paula was one of them, she never wanted to have children," he reminded me.

"It's possible she feared the consequences of the por-phyria, not just the risk for her but because she could transmit it to her children."

"Long before she suspected she had porphyria, my sister said that children are adorable only at a distance, and that there are ways to fulfill oneself other than by being a mother. There are also women in whom the maternal instinct is never awakened. If they get preg-nant, they feel they're invaded by a strange being that is consuming them, and then they don't want the child. Can you imagine the scar that's left in the soul of some-one who is rejected at birth?"

"It's true, Nico, there are exceptions, but by far the largest majority of women want children and when they do have them sacrifice their lives for them. There's no danger that humanity will fade out for lack of children."

Mail-Order Bride

Lili came from China on a prospective bride's visa that was good for three months, at the end of which she had to marry Tong or return to her country. She was a pretty, healthy woman who looked about twenty, though she was almost thirty, and she was very little contaminated by occidental culture, just as her future husband wished. She didn't speak a single word of English; so much the better, for that would make it easier to keep her submissive, was the opinion of her future mother-in-law, who from the first applied ancient traditions in making her daughter-in-law's life impossible. We found her moon face and sparkling eyes irresistible; even my grandchildren fell in love with her. "Poor girl, it's going to be difficult for her to adapt," Willie commented when he learned that Lili got up at dawn to do

the housework and prepare the complicated dishes demanded by the old woman, who in spite of her minuscule size pushed Lili around. "Why don't you tell the
old bitch to go to hell?" I asked Lili with signs, but she
didn't understand. "Keep your nose out of it," Willie
recited once again, and added that I knew nothing about
Chinese culture . . . but I knew more than he did, I had
at least read Amy Tan. The mail-order bride was not as
fainthearted as Willie had reported when he met her,
of that I was sure. She had a peasant stolidity, broad
shoulders, determination in her look and actions; with a
flick of her wrist she could break Tong's mother's neck,
and his too, if she chose. No sweet little dove there.

After three months, when Lili's visa was about to
expire, Tong told us that they were getting married.
Willie, as a lawyer and friend, reminded him that the
girl's only reason for marrying was to stay in the United
States. She would need a husband only two years; after
that she could divorce and get her residence permit.
Tong had thought about it; he wasn't so naive as to
believe that a girl off the Internet had fallen in love
with him when she saw his photograph, however much
Lori had retouched it; he believed, however, that both
of them had something to gain from the arrangement:
he, the possibility of a son, and she, a visa. They would
see which of the two came first, it was worth the risk.

Willie advised him to have a prenuptial contract drawn up, otherwise she would be entitled to part of the savings he had accumulated with such sacrifice, but Lili said she would not sign a document she couldn't read. They went to a lawyer in Chinatown, who translated it. When she realized what Tong was asking of her, Lili turned the color of a beet and for the first time raised her voice. How could they accuse her of marrying for a visa! She had come to make a home with Tong! she protested, immersing the groom and the lawyer in a flood of repentance. They were married without the agreement. When Willie told me about it, sparks were shooting out of his ears; he couldn't believe that his bookkeeper was so dumb; what was making him do such a stupid thing? he was fucked for good now; couldn't Tong remember how he, Willie, had been fleeced by every woman who passed by? and on and on with a litany of gloomy prognoses. For once, I had the pleasure of getting back at him: "Keep your nose out of it."

Lili enrolled in an intensive English class and wore headphones all day, listening to the lesson until she fell asleep, but her apprenticeship was slower and more difficult than she'd expected. She went out to look for a job, but in spite of her hard-earned education and her experience as a nurse she couldn't find anything because she didn't speak English. We asked her to clean our house and pick up the children at school because by then Ligia wasn't working for us anymore. One by one she had brought her children from Nicaragua and put them through school, and now they were all professionals. At last she could rest. If Lili was working for us, she could earn a decent salary until she found something appropriate for her skills. She gratefully accepted, as if we had done her a favor, when she was the one who was helping us.

At first, communication with Lili was amusing: I left drawings fastened to the refrigerator, but Willie's method was to shout at her in English, to which she answered "No!" with an adorable smile. Once Roberta came to visit; she is a transsexual friend who before she was a woman had been an officer in the Marines named Robert. He fought in Vietnam, was decorated for courage, was horrified at the death of innocents, and left the military service. For thirty years he lived with his wife, whom he loved and who was his companion during the process of his becoming a woman, and they stayed together until she died of breast cancer. To judge by the photographs, Roberta had been a hefty, hairy man with a broken nose and the chin of a corsair. He had gone through hormone treatments, plastic surgery, electrolysis to remove his facial hair, and finally an operation on his genitals, but I suppose the result still was not convincing, because Lili stood staring at Roberta openmouthed, and then took Willie behind a door to ask him something in Chinese. My husband deduced that it was about our friend's gender, and began to explain to Lili in a whisper that kept getting louder and louder until he ended up yelling at the top of his lungs that Roberta was a man with the soul of a woman, or something like that. I nearly died of embarrassment, but Roberta kept drinking tea and eating little pastries with her beautiful manners, ignoring the shouting behind the door.

My grandchildren and Olivia, the dog, adopted Lili. Our house had never been so clean; she disinfected it as if planning open heart surgery in the dining room. Gradually she was incorporated into our tribe. When she married, her shyness vanished; she took a deep breath, stuck out her chest, got a driver's license, and bought a car. She brightened Tong's life. He is even better looking because Lili dresses him with style and cuts his hair, though that doesn't mean there aren't sparring matches; he is a despotic husband. I tried to mime to her that the next time he raised his voice to her, she should crack him over the head with a skillet, but I don't think she understood. All that's missing is children, which don't come along because she has fertility problems and he isn't young anymore. I suggested they adopt in China, but they don't give away boys there, and "Who wants a girl?" The same words I heard in India.

Magic for the Grandchildren

When I finished *Portrait in Sepia*, I was troubled by a promise that I could not keep postponing, which was to write three adventure novels for Alejandro, Andrea, and Nicole, one for each of them. As I had done with my children, I told my grandchildren stories from the time they were born, following a system we had refined to perfection: they would give me three words, or three subjects, and I had ten seconds to invent a story that would use all three. They plotted together to give me the most nonsensical cues possible, and bet that I wouldn't be able to weave them together, but my training—it had begun in 1963 with you, Paula—was as formidable as their innocence, and I never failed them. The problem would come when, for example, they asked me to repeat word for word the story about a restless ant that fell into an inkwell and accidentally discovered Egyptian writing. I didn't have the slightest recollection of that erudite insect and found myself in trouble when they suggested that I consult my mental computer. "The life of ants is a bore, nothing but work and serving the queen; I'd rather tell you the story of a murderous scorpion," and I launched into that tale before they had time to react. But a day came when not even that dodge worked, and that was when I promised I would write three books on subjects they proposed, just as we had done with the improvised-in-ten-seconds bedtime stories.

My grandchildren gave me the theme for the first book, which I had already sensed in many of the stories they had asked me for: ecology. The adventure of *The City of the Beasts* evolved from the trip to the Amazon. Now I know that when my well of inspiration dries up, as happened following your death, Paula, I can refill it by taking trips. My imagination awakens when I leave my familiar surroundings and confront other ways of life, different people, languages I don't command, when I'm exposed to unforeseen vicissitudes. I can tell that the well is filling because my dreams become more active. The images and stories I accumulate on the trip are transformed into vivid dreams, sometimes into violent nightmares, that announce the arrival of the muses. In the Amazon I sank into a voracious nature, green on green, water on water; I saw caimans the size of a rowboat, pink dolphins, manta rays floating like carpets in the tea-colored waters of the Río Negro, piranhas, monkeys, unbelievable birds, and snakes of assorted varieties, including an anaconda—dead, but an anaconda nevertheless. I thought I'd never use any of that because it didn't fit into the kinds of books I write, but it all turned out to be useful when it was time to plot the first juvenile novel. Alejandro was the model for Alexander Cold, the protagonist; his friend Nadia Santos is a blend of Andrea and Nicole. In the novel, Alexander accompanies his grandmother Kate, a travel writer, to the Amazon, where he meets Nadia. The young people get lost in the jungle, live with a tribe of "invisible Indians," and discover some prehistoric beasts that live inside a *tepuy*, the strange geological formation of the region. The idea of the beasts emerged from a conversation I overheard in a restaurant in Manaus among a group of scientists who were commenting on the find in the jungle of a gigantic fossil that had human aspects. They were wondering what animal family it corresponded to; maybe it belonged to the family of the monkeys or was a kind of tropical yeti. With those facts it was easy to imagine the beasts. Invisible Indians really do exist; they are tribes that live in the Stone Age and that in order to blend into their surroundings paint their bodies, imitating the vegetation around them, and move so stealthily that they can be ten feet away and not be seen. Many of the stories I heard in the Amazon about corruption, greed, illegal trafficking, violence, and smuggling were raw material for the plot. What was essential, however, was the jungle, which became the setting and determined the tone of the book.

A few weeks after beginning the first volume of the trilogy, I realized that I was incapable of the flights of imagination the project required. It was very difficult for me to crawl into the skin of those two teenagers who would live a wondrous adventure aided by their "spirit animals," as is the tradition of some indigenous tribes. I recall the terrors of my own childhood, when I had no control over my life or the world around me. I was afraid of very specific things: that my father, who had disappeared so many years ago that even his name had been lost, would come to reclaim me, or that my mother would die and I would end up in some gloomy orphanage eating cabbage soup; but most of all I was frightened by the creatures that peopled my own mind. I thought that the devil appeared at night in the mirrors; that the dead came out of the cemetery during earthquakes, which in Chile are very common; that there were vampires in the attic, large evil toads in the armoires, and souls in pain in the sitting room curtains; that our neighbor was a witch and the rust in the pipes was the blood of human sacrifice. I was sure that the ghost of my grandmother sent me cryptic messages in the bread crumbs or in the shapes of clouds, but that didn't frighten me, it was one of my few calming fantasies. The memory of that ethereal and entertaining grandmother has always been a consolation, even now that I am twenty-five years older than she was when she died. Why wasn't I encircled by fairies with dragon-fly wings or sirens with bejeweled tails? Why was everything so horrible? I wouldn't know; maybe most children live with one foot in those nightmarish universes. To write my novels for youthful readers I couldn't call upon the macabre fantasies of those years, since that wasn't so much a case of evoking them as it was of feeling them in my bones, the way you do in childhood, with all their emotional charge. I needed to be again the little girl I once was, that silent girl tortured by her own imagination, who wandered like a shadow through her grandfather's house. I had to demolish my rational defenses and open my mind and heart. And to do that I decided to subject myself to the shamanic experience of *ayahuasca,* a brew the Amazon Indians prepare from the climbing plant *Banisteriopsis* to produce visions.

Willie did not want me to take that trip alone, and as on so many occasions of our shared lives, he blindly accompanied me. We drank the dark, foul-tasting tea, barely a third of a cup but so bitter and fetid it was nearly impossible to get down. It may be that I have a flawed cerebral cortex—I am always a little off the ground—because the *ayahuasca*, which gives others a push toward the world of the spirits, catapulted me so far that I didn't come back till a couple of days later. Within fifteen minutes of taking it, I lost my balance and curled up on the floor, unable to move from there. I was panicked, and called to Willie, who was able to drag himself to my side, and I clung to his hand as I would to a life buoy in the worst storm imaginable. I couldn't talk or open my eyes. I was lost in a whirlpool of geometric figures and brilliant colors, which at first were fascinating and then exhausting. I felt that I was leaving my body, that my heart was bursting, and I was filled with a terrible anguish.

Soon colors vanished and the black rock appeared that normally lay nearly forgotten in my chest, as threatening as some Bolivian mountains. I knew that I had to remove it from my path or I would die. I tried to climb over it but it was slippery; I wanted to go around it but it was too large; I began to tear pieces from it but the task was unending; and all the while

my certainty was growing stronger that the rock contained all the world's evil; it was filled with demons.
I can't guess how long I was there; in that state time
had nothing to do with the time we're accustomed to.
Suddenly I was shaken by an electric charge of energy.
I gave a formidable kick and leaped atop the rock. For
a moment I returned to my body, doubled over with
nausea. I felt for the basin I had left within reach and
vomited bile. Nausea, thirst, sand in my mouth, paral-
ysis. I perceived, or understood, what my grandmother
used to tell me, that space is filled with presences
and that everything happens simultaneously. Images
were superimposed on images, transparent, like those
illustrations on clear pages in science texts. I wandered
through gardens where threatening plants with flesh-
eating leaves were growing, large mushrooms that
were oozing poison, malevolent flowers. I saw a little
four-year-old girl, shrinking back, terrified; I held
out my hand to pull her up, and it was me. Different
periods and persons passed from one illustrated page
to another. I found myself with myself in different
moments and in other lives. I met an old gray-haired
woman, small but erect, with gleaming eyes; she, too,
could have been me, a few years into the future, but
I'm not sure because she was surrounded by a milling
multitude.

Soon that populated universe vanished and I entered a white, silent space. I was floating on air; I was an eagle with its great wings outspread, riding the wind, seeing the world from above, free, powerful, solitary, strong, unbound. That great bird was there for a long time, and then suddenly it soared to a different, still more glorious place where form disappeared and there was nothing but spirit. There was no eagle, no memories, no emotions, there was no "I," I had dissolved in the silence. If I'd had the slightest awareness or desire, I would have looked for you, Paula. Much later, I saw a small circle like a silver coin, and shot toward it like an arrow; I crossed through the opening and effortlessly plunged into an absolute void, a deep translucent gray. There was no sensation, no spirit, not a trace of individual consciousness; instead I felt a divine, absolute presence. I was inside the Goddess. It was the death or glory the prophets speak of. If that was dying, Paula, you are in a dimension beyond reach, and it is absurd to imagine that you are with me in everyday life, or that you are helping me in my tasks and ambitions, my fears and vanities.

A thousand years later I returned, like an exhausted pilgrim, to a familiar reality, following the same road I had taken there, but in reverse: I went through the small silver moon, floated in the eagle's space, descended to the white sky, sank into psychedelic images, and finally reentered my poor body, which for two days had been very ill, cared for by Willie, who was beginning to believe he had lost his wife to the world of the spirits. In Willie's reaction to the *ayahuasca*, he did not ascend to glory nor enter death, he was trapped in a bureaucratic purgatory, shuffling papers, and a few hours later the effects of the drug had passed. In the meantime, I was lying on the floor, where he later made me comfortable with pillows and blankets, shivering, muttering incomprehensible words, and vomiting a foam that was whiter with each retching. At first I was agitated, but later I lay relaxed and motionless; I didn't seem to be suffering, Willie tells me. The third day, by then conscious, I spent in bed reliving each instant of that extraordinary journey. I knew now that I could write the trilogy, because to counteract a stumbling imagination I had the opportunity to perceive the universe, once again, with an intensity provided by the *ayahuasca*, similar to the fervor of my childhood. The adventure with the drug bound me with something I can only define as love, an impression of oneness: I dissolved into the divine, I felt that there was no separation between me and the rest of all that exists, all that was light and silence. I was left with the certainty that we are spirits, and all that is material is illusory, something that cannot be proved rationally but at times I have briefly experienced in moments of exaltation before nature, of intimacy with a beloved, or in meditation. I accepted that in this human life my totemic animal is the eagle, the bird that glided through my visions viewing everything from a great distance. That distance is what allows me to tell stories, because I can see angles and horizons. It seems that I was born to tell and tell and tell. My body ached, but I have never been more lucid. Of all the adventures in a lifetime of upheaval, the only thing I can compare to that visit to the dimension of the shamans was your death, daughter. On both occasions something inexplicable and profound happened that transformed me. I was never the same after your last night in this world, or after I drank that powerful potion: I lost my fear of death and experienced the eternity of the spirit.

Empire of Terror

On Tuesday, September 11, 2001, I was in the shower when the telephone rang early in the morning. It was my mother, from Chile, horrified at the news we hadn't as yet heard because it was three hours earlier in California than on the other coast and we'd just got out of bed. When I heard her voice I thought she was talking about the anniversary of the military coup in Chile; it, too, was a terrorist attack against a democracy, which we remember every year as a day of mourning: Tuesday, September 11, 1973. We turned on the television and watched over and over the same images of the planes crashing into the towers of the World Trade Center that reminded me of the bombardment the military launched against La Moneda palace in Chile, the place where President Salvador Allende died that day. We ran to our banks to withdraw cash and to get in a supply of water, gasoline, and food. Flights were canceled, thousands of passengers were trapped, hotels were booked beyond capacity and had to put beds in the hallways. Telephone lines were so overloaded that communication was nearly impossible. Lori couldn't reach her parents for two days, and I wasn't able to talk with mine in Chile. Nico and Lori moved over to our house with the children, who were with them that week; they didn't go to school because classes were canceled. Together we felt more safe.

For days, no one could go back to work in Manhattan; a cloud of dust floated in the sky and toxic gases escaped from broken pipes. In the midst of the still reigning confusion, we had news from Jason. He told us that in New York the situation was slowly beginning to improve. He walked at night to the area of the disaster with a spade and helmet to help the rescue teams, which were exhausted. He passed dozens of volunteers returning from hours of labor in the ruins with white cloths tied around their necks in honor of the victims trapped in the towers, who had waved handkerchiefs out the windows to say good-bye. From a long way away you could see smoke rising from the ruins. New Yorkers felt as if they'd been clubbed. Sirens sounded and ambulances rolled by empty because there were no further survivors, while dozens of television cameras lined up around the area marked out by the firemen. Everyone was anticipating another strike, but no one seriously considered leaving the city; New York had not lost its ambitious, strong, visionary character. When Jason reached the site of the disaster, he met hundreds of volunteers like himself; for each victim that disappeared in the ruins there were several persons ready to look for him. Every time a truckload of workers drove by, the crowd greeted them with shouts of encouragement. Other volunteers brought water and food. Where once proud towers had stood there was a black, smoking hole. It's like a terrible dream, Jason told us.

It wasn't long until the bombing of Afghanistan began. Missiles rained over the mountains where the handful of terrorists were hiding—no one wanted to confront them face to face—their concussions leveling the earth, while winter fell over Afghanistan and women and children in the refugee camps began to die of the cold: collateral damage. In the meantime paranoia was growing in the United States; people were wearing gloves and masks to open mail, fearing the possibility of a smallpox virus or anthrax, supposed weapons of mass destruction. As terrified as anyone, I went out and bought Cipro, a powerful antibiotic that could save my grandchildren in case of biological warfare, but Nico told me that if we gave the pill to the children at the first symptom of a cold, it wouldn't be effective for a real illness. It was like killing flies with a cannon. "Be calm, Mamá, you can't prevent everything," he told me. And then I remembered you, daughter, and the military coup in Chile, and many other moments in my life when I was powerless. I have no control over crucial events, those that determine the course of life, so it makes more sense to relax. The collective hysteria had made me forget that essential lesson for several weeks, but Nico's comment restored my sense of reality.

Juliette and the Greek Boys

During the course of my research for the trilogy for young readers, I often visited the Book Passage bookstore, and it was there I met Juliette, a young, very beautiful, and very pregnant American girl who was barely managing to counterbalance the most enormous belly I had ever seen. She was expecting twins, but they were not hers, she told me; they belonged to someone else, and she had merely lent her womb. It was an altruistic impulse on her part, which, after I learned her story, seemed truly foolish.

At the age of twenty-something, after her university graduation, Juliette took a trip to Greece, a logical destination for someone who had studied art. There she met Manoli, an exuberant Greek with a mane of hair and a prophet's beard, velvety eyes, and an overpowering personality that immediately seduced her. The man wore very short shorts, and when he crouched down or sat and crossed his legs, his private parts were no longer private. I can imagine they were exceptional, since women chased him at a fast trot through all the little streets on the island. Manoli had a silver tongue and could spend twelve hours in the plaza or in a café telling stories without taking a breath, surrounded with listeners hypnotized by his voice. The story of his own family was a novel in itself: the Turks had decapitated his grandfather and grandmother before their seven children, who, along with other Greek prisoners, were forced to walk from the Black Sea to Lebanon. Along that route of sorrow six of the siblings died and only Manoli's father, who was then six years old, survived. Among the hundreds of tourists golden from the sun and eager to roll with him on the warm Grecian sands, Manoli chose Juliette, for her air of innocence and her beauty. To the surprise of the island's inhabitants, who considered him to be an incorrigible bachelor, he proposed marriage to Juliette. There had been a previous marriage to a Chilean woman who, bizarrely, had run away with a yoga instructor on the day of her wedding. The story wasn't clear, but according to local gossip, the rival had put LSD in Manoli's drink, and he had awakened a day later in a psychiatric hospital. By then his scatterbrained wife had disappeared. He never heard anything of the Chilean woman again, and in order to marry Juliette he had to cut through the red tape to prove that his wife had deserted the marriage, since there had been no one to sign the divorce papers.

Manoli lived in an old house atop a cliff overlooking the Aegean sea; for more than two hundred years it had belonged to a succession of lookouts whose responsibility it was to scan the horizon. At the sight of enemy ships, they had to jump on a horse, kept always saddled, and gallop to the mythic city of Rhodes, founded by the gods, to sound the alarm. Manoli set tables outside and converted it into a restaurant. Every year he put a coat of white paint on the house and dark brown on the shutters and doors, like all the homes in that idyllic town where there are no cars and people know each other by name. Lindos, crowned by its acropolis, looks more or less as it has for many centuries, with the addition of a medieval castle, now in ruins. Juliette didn't hesitate an instant to marry, although she knew from the beginning that this was a man who could never be tamed. To avoid the pain of jealousy and the humiliation of having someone come to her with the latest gossip, she informed Manoli that he could have all the amorous adventures he pleased, but never behind her back, she would rather know. Manoli thanked her, but fortunately he had the good sense never to confess an infidelity, and as a result Juliette lived in peace and in love. She and Manoli were together for sixteen years in Lindos.

The restaurant kept them very busy during high season, but it closed in winter and they used that time to travel. Manoli was a magician in the kitchen. He prepared everything at the moment: meat, grilled fish, fresh salads. He himself chose each fish the boats brought from the sea at dawn, and every vegetable that came on mule back from the gardens outside of town, and his fame spread beyond the island. It was a twenty-minute stroll from the town to the cliff where the restaurant stood. The clientele was in no hurry; the majestic countryside invited contemplation. Most stayed through the night to follow the trajectory of the moon above the acropolis and the sea. Juliette, with her classic face, her light cotton dresses, her sandals, and her dark chestnut hair loose on her shoulders, was even more attractive than the food. She looked like a vestal virgin of some ancient Greek temple, and it came as a shock to hear her American accent. She glided among the tables with her trays, always sweet and pleasant despite the tumult of the customers crowded around the tables and awaiting their turn at the door. Only twice did she lose patience, and in both instances it was with American tourists. The first had to do with a blimp of a man, red-faced from too much sun and ouzo, who three times sent his plate back because it was not precisely what he wanted. Worse, he did so with objectionable manners. Juliette, exhausted after a long night of serving customers, brought him the fourth plate and without a word dumped it over his head. The second occasion was the fault of a snake that curled up a table leg and slithered toward the salad bowl in the midst of hysterical screams from a group of Texans who undoubtedly had seen others much larger where they came from. Juliette saw no reason to frighten the customers with that uproar. She fetched a large knife from the kitchen and with four karate chops cut the snake into five neat pieces. "I'll be right out with your lobster," was all she said.

Juliette willingly put up with Manoli's manias—he was never an easy husband—because he was the most entertaining and passionate man she had ever known. Compared to him, other men seemed insignificant. Women, right in front of her, handed Manoli the key to their hotel rooms, which he always refused with some charming joke—after carefully taking note of the room number. They had two boys as good-looking as their mother: Aristotelis, and then four years later, Achilleas. The younger was still in diapers when his father went to Thessaloniki to consult a physician about his aching bones. Juliette stayed in Lindos with the boys, looking after the restaurant as best she could, not attaching too much significance to her husband's aches and pains since she had never heard him complain. Manoli phoned every day to talk about trifles, never referring to his health. If she asked a question, he answered evasively, with the promise that he would be back before the week was out, when they learned the results of the tests. However, the very day she was expecting him back, about dusk, she saw a long line of friends and neighbors coming up the hill toward her house. She felt a claw in her throat and instantly recalled that the day before, during his phone call, her husband's voice had cracked with a sob when he told her, "You are a good mother, Juliette." She had been thinking about those words, so unexpected from Manoli, who was not given to heartfelt compliments. At that moment she realized that it had been his good-bye. The sorrowful faces of the men gathered at her door, and the collective embrace of the women, confirmed her fears. Manoli had died of a galloping cancer that no one had suspected because he had been so clever in hiding the torment of his deteriorating bones. He had gone into the hospital knowing that his hour had come, but out of pride had not wanted his wife and sons to see him die in agony. Juliette's neighbors coordinated their efforts and bought plane tickets for her and her boys. The women packed her suitcase, closed the house and restaurant, and one of them went with her to Thessaloniki.

The young widow went from one hospital to the next, looking for her husband; she wasn't even sure where he could be found, but finally she was led to a cellar, no more than a cave in the earth, like those used for storing wine, where a corpse lay on a slab, barely covered by a sheet. Her first feeling was one of relief; she thought she had been the victim of a terrible mistake. That yellow, skeletal cadaver wearing a twisted expression of suffering did not remotely resemble the happy, full-of-life man who was her husband. But then the orderly who had showed her the way held the lamp close, and Juliette recognized Manoli. In the hours that followed, Juliette had to dredge up strength, find a cemetery plot, and bury her husband without ceremony. Afterward, she took her sons to a nearby plaza, and there amid the pigeons and the trees, she explained that they would not see their father again but that they would often feel him at their side; Manoli would always look after them. Achilleas was too young to understand the immensity of his loss, but Aristotelis was terrified. That same night Juliette awakened with a start, certain that she was being kissed. She felt soft lips, a warm breath, the tickle of her husband's beard. He had come to give her the farewell kiss he had not wanted to give before, when he was dying alone in the hospital. What she had told her children to console them was absolute truth: Manoli would look after his family.

The town of Lindos closed ranks around the young widow and her children, but their embrace could not sustain her indefinitely. Juliette could not manage the restaurant alone, and since she couldn't find other work on the island, she decided that the moment had come to rejoin her family in California, where at least she could count on her parents' help. Life changed for the boys, who had been brought up free and secure, playing barefoot in the white streets of Lindos, where everyone knew them. Juliette found a modest apartment, part of a church project, and was hired by Book Passage. She had no more than moved in when her mother was diagnosed with a terminal illness, and after only a few months Juliette had to bury her. One year later her father died. There had been so much death around her that when she heard of a couple who were looking for a surrogate womb in which their child might grow, she offered hers without much thought, with the hope that the life within her would console her for so many losses, and give her warmth. I met Juliette when she was deformed by the pregnancy: swollen legs, blotches on her face, circles under her eyes . . . completely exhausted, but happy. She kept working in the bookstore until she was forced to stop by order of her physician; she spent the last weeks on a sofa, crushed by the weight of her belly. In fewer than four years Aristotelis and Achilleas had lost their father and two grandparents; their short lives were marked by death. They clung to their mother, the one person they had left, with the inevitable fear that she, too, might disappear. For that very reason it seemed strange to me that Juliette would run the risk of that pregnancy.

"And who will the parents of these twins be?" I asked her.

"I scarcely know them. The contact was made through a bereavement group I meet with every week, adults and children going through a period of pain. The group has helped us a lot; now Aristotelis and Achilleas understand that they are not the only ones who don't have a father."

"The agreement with that couple was that you would have one baby, not two. Why should they get a bonus? Give them one baby and hand the other to me."

Juliette burst out laughing, and explained that neither of them belonged to her; there were strict rules, even legal contracts, in regard to eggs, sperm, and paternity, so I couldn't take one of the twins for myself. A shame, but it wasn't the same as a litter of pups.

Juliette is the goddess Aphrodite, all sweetness and abundance: curves, breasts, kissable lips. Had I known her earlier, her image would have graced the cover of my book about food and love. It seemed normal that she and her two Greek boys, as we call her sons, would become a part of our family, and now when I count my grandchildren, I have to add two more. So the tribe is growing, this blest group in which happiness is multiplied and sorrow divided. The most prestigious private school in the county offered scholarships to Aristotelis and Achilleas, and by a stroke of luck, Juliette was able to rent a small house with a patio and garden in our neighborhood. Now all of us, Nico, Lori, Ernesto, Giulia, Juliette, and Willie and I, live within the radius of a few blocks, and the children can go from one house to another either on their bikes or walking. The family helped Juliette move, and while Nico was making some repairs, Lori was hanging pictures, and Willie was installing a grill, I was calling on Manoli to look after his own from the other side, just as he had promised with the posthumous kiss he gave Juliette in farewell.

One summer afternoon we were all sitting around the pool at our house watching Willie teach Achilleas to swim—he was afraid of the water but he was green with envy when he saw the other children splashing around—and I asked Juliette how she, such a maternal person, could carry two babies for nine months, give birth to them, and then that same day let them be taken away.

"They were never mine. They were in my body for a while, that was all. While I was carrying them I cared for them and loved them tenderly, but it wasn't the possessive love I feel for Aristotelis and Achilleas. I always knew we would be separated. When they were born, I held them for a moment in my arms. I kissed them and wished them good luck in life, and then I handed them to their parents, who immediately took them away. My breasts, heavy with milk, ached afterward, but not my heart. I was happy for that couple who so badly wanted children."

"Would you do it again?"

"No. I'm nearly forty years old, and pregnancy takes a lot out of you. I would only do it for you, Isabel," she told me.

"For me? God forbid! The last thing I want at my age is an infant." I laughed.

"Then why did you ask if you could steal one of the twins for yourself?"

"It wasn't for me, it was for Lori."

Jason and Judy

In my mother's eyes, Willie's best quality is that he is "well trained." It would never have occurred to her to telephone Tío Ramón at the office to ask him to pick up sardines for dinner, or ask him to take off his shoes, climb on a chair, and go over the top of some piece of furniture with the feather duster, things that Willie does without a fuss. As for me, I think my husband's most admirable quality is his stubborn optimism. There is no way to sink Willie. I have a few times seen him on his knees, but he gets up, brushes off the dust, claps his hat on his head, and keeps going. He has had so many problems with his children that had I been in his place I would have been incurably depressed. It wasn't just Jennifer he suffered over; his two sons have lived dramatic lives owing to their addiction to drugs. Willie has always helped them, but with the passing of the years his hopes have been flagging; that is one reason he clings to Jason.

"Why are you the only one who learned something from me? All the others do is ask: give me, give me, give me," Willie told him once.

"They feel entitled because they're your sons, but you don't owe me anything. You aren't my father but you've always looked after me. Why wouldn't I think what you tell me is important?" Jason replied.

"I'm proud of you," Willie grunted, disguising a smile.

"That's not too difficult, Willie, your yardstick isn't very high."

Jason adapted to New York, the most entertaining city in the world, where he is successful in his work; he has friends there and makes a living with his writing, and he has found the girl, "as reliable as Willie," he was looking for. Judy graduated from Harvard, and writes on sex and relationships for the Internet and women's magazines. She has a Korean mother and North American father; she's beautiful, intelligent, and as fiercely independent as I am. She can't tolerate the idea that someone is supporting her, partly because she saw her mother—who barely speaks English—completely subjected by her father, who in due time left her for a younger woman. Judy had cured Jason of his vice of exploiting his drama as a way to seduce women. With the story of the girlfriend who left him for his sister-in-law, Jason had all the dates he wanted; he never lacked for a female shoulder, and a little more, when he sought consolation. With Judy that formula didn't work; she learned early to make her own way, and she is not a person to complain. She was sympathetic to what he had gone through, but that was not what had attracted her. When Jason met her she had been living with another man for four years, but she wasn't happy.

"Do you love him?" Jason asked her.

"I don't know."

"If it's that difficult to answer that question, it's probably because you don't love him."

"What do you know? You don't have any right to say that," she replied indignantly.

They kissed, but Jason told her that they wouldn't even touch until she left the fellow because he wasn't inclined to be dumped another time. In less than a week she left the stupendous apartment where she was living—which seems to be the ultimate test of love in New York—and moved to a dark garret at some distance from the center of the city. A couple of years went by before the relationship jelled because Jason still didn't trust women in general and marriage in particular, since his parents, stepmothers, and stepfathers had been divorced one, two, even three times. One day Judy told him that she wasn't going to be made to pay for Sally's betrayal. That, plus the fact that she loved him even though he was resistant to making a commitment, brought a reaction. Finally he could let down his defenses and laugh about the past. Now from time to time he even communicates with Sally by e-mail. "I'm happy she's been with Celia for so long, that means that she didn't leave me on a whim. A lot of people were hurt but at least something good has come out of all this mess," he told me.

According to Jason, Judy is the most decent person he knows, without the least affectation or malice. She is always surprised by the cruelty of the world because it would never occur to her to harm anyone. She adores animals. When they met she was walking abandoned dogs with the hope that someone would be attracted to them. At the time she was walking Toby, a pathetic animal that looked like a hairless mouse; he urinated uncontrollably and suffered attacks of epilepsy. He would lie on his back with four stiff paws in the air, foaming at the mouth. He was the fourth dog she'd taken care of, but there was no hope that anyone would ever fall in love with such a horror and take him home, so she took him to Jason, to keep him company as he wrote. In the end, they kept poor Toby.

For more than a year, Jason had been working for a men's magazine, one of those with full-color pages of lascivious girls with open lips and open legs, when he was given an assignment to cover the strange crime of a young man who had killed his best friend in the New Mexico desert, where they'd gone to camp. They were lost, and close to dying when one man asked the other to give him a merciful death, he didn't want to die of thirst . . . and his friend killed him with his knife. The circumstances were rather murky, but the judge determined that the murderer was temporarily insane from dehydration, and let him go with a minimal penalty. Jason's assignment was not an easy one because despite the notoriety of the crime; the details had not been spelled out during the trial, and neither the accused nor his friends and family would speak with him. He had to base his report on what he gathered at the site of the crime and the commentaries of rangers and police. Nevertheless, even with so little material, he gave his article the urgency and suspense of a detective novel. A week after the magazine hit the street, a publishing house contracted him to write a book on the case, and paid him an advance unusual for a new author. It was published with the title *Journal of the Dead*. Soon it fell into the hands of some film producers, and Jason sold the rights to the movie. Overnight he was on his way to

becoming the next Truman Capote. He moved easily
from journalism to literature, just as I had predicted
the first time he showed me one of his stories, when he
was eighteen and vegetating in Willie's house wrapped
in a blanket and drinking beer at four in the afternoon.
That was the period in which he didn't want to separate
from the family and called our office at mid-afternoon
to ask us what time we would be back home, and what
were we going to fix him for dinner. Now he is the only
one of our brood who doesn't need any help. With the
money from the book and the movie, he decided to buy
an apartment in Brooklyn. Judy suggested that they
each pay half, and to the stupefaction of Jason and the
rest of the family, she wrote a check for six figures. She
had worked hard since she was a teenager, and she is
frugal and knows how to invest her money. Jason won a
prize with that girl, but she doesn't want to marry him
until he stops smoking.

Buddhist Mothers

Fu and Grace had not adopted Sabrina, it hadn't seemed crucial, but then Jennifer's old boyfriend got out of prison, where he'd been serving time, and made clear his intent to see his daughter. He had never agreed to have a blood test to prove his paternity, and in any case he had lost his rights as a father, but his voice on the telephone put them on the alert. The man wanted to have the little girl on weekends, something the mothers did not want at all because of his police record and a way of life that they had little confidence in. They decided that the moment had come to make the situation with Sabrina legal. That coincided with the death of Grace's seventy-five-year-old father, who had smoked his entire life; his lungs were destroyed and he had ended up in a hospital connected to a respirator. He lived in Oregon, the only state in the country where no one invokes the law when a terminally ill person chooses the moment he wants to die. Grace's father had figured that to go on living in that terrible condition would cost a fortune, and it wasn't worth it. He called his children, who gathered from around the country, and using his laptop explained that he had summoned them to tell them good-bye.

"Where are you going, Father?"

"To heaven, if they'll let me in," he wrote on the screen.

"And when are you planning to die?" they asked, amused.

"What time is it?" the patient wanted to know.

"Ten o'clock."

"Let's say about noon," he wrote. "How does that seem?"

And at exactly noon, after saying good-bye to each of his astounded descendants and consoling them with the idea that this solution was best for all of them—especially him, because he wasn't planning to spend years hooked up to a respirator, and besides, he had a burning curiosity to see what lay on the other side—he disconnected the respirator and died happy.

For Sabrina's adoption, a female judge came from San Francisco, and we appeared before her in a family group. From the door of a chamber in City Hall, we could see coming toward us down a long corridor that miraculous granddaughter walking for the first time without the help of a walker. Her small figure advanced with difficulty along that endless, tiled hallway, followed by her mothers, right behind her but not touching, ready to intervene if it were necessary. "Didn't I tell you I was going to walk?" Sabrina said defiantly, with the touch of pride that crowns each tenacious conquest. She was wearing a party dress and pink slippers, with ribbons in her hair. She told us all hello, ignoring Willie's emotion, posed for photos, thanked us for being there, and solemnly announced that from that moment her name was Sabrina, plus Jennifer's surname, followed by those of her adopting mothers. Then she turned to the judge and added, "The next time we see each other, I will be a famous actress." And we were all convinced that she would be. Sabrina, raised in the macrobiotic, spiritual Zen Center retreat, aspires to be a movie star, and her first choice in food is a bloody hamburger. I don't know how, but she is invited every year to the Academy Awards ceremony in Hollywood. On Oscar night, we see her on television, seated in the gallery with a notebook in hand to keep count of the celebrities that parade down the red carpet. She's in training for the time when it's her turn to walk down that same red carpet.

Fu and Grace are not a couple any longer, after having been together for more than a decade, but they are still united through Sabrina and a friendship of such long standing that it doesn't make sense to separate. They rearranged the little dollhouse they have at the center; as small as it is, it is greatly coveted because there are always postulants eager to live in that calm pool of spirituality. They divided the space, left one room in the middle for Sabrina, and live at either end. You have to jump over furniture and scattered toys in those tiny rooms, which they also share with Mack, one of those big canines trained to be a service dog. They acquired him for Sabrina and she loves him very much, but she doesn't need him; she can navigate on her own. It took a year of rigorous negotiations to obtain Mack; they had to take a course in how to communicate with him, they were sent an album containing photos of him as a pup, and they were warned that they would have surprise visits from an inspector, and if they were not caring for him properly, he would be taken from them. Finally he arrived: an off-white Labrador with eyes like grapes, sharper than most humans. One day Grace took him with her to the hospital and he followed on her rounds; she noticed that even her dying patients perked up in Mack's presence. She had a psychotic patient who'd been sunk in his personal purgatory for a very long while; he had a deformed hand he always kept hidden in a pocket. Mack entered his room wagging his tail; gently he rested his enormous head on the man's knees and sniffed and nosed his pocket until the man pulled out the hand he was so embarrassed about, and Mack began to lick it. Perhaps no one had ever touched him that way. The sick man's eyes met Grace's, and for an instant it seemed that he had come out of the dark cell in which he was trapped into the light. From then on, Mack has been kept busy at the hospital; they hang a placard labeled "Volunteer" on his chest and send him on his rounds. The patients hide the cookies from their meals to give to him, and Mack has become rather portly. Compared to that animal, my Olivia is no more than a mop of hair with the brain of a fly.

While Grace and Mack are working at the hospital, Fu is in charge of the Zen Center, where one day she will be abbess, though she has never evidenced any interest in the post. That imposing woman with the shaved head and dark robes of a Japanese monk always produces the impact I felt the first time I saw her. But Fu is not the only notable woman in her family. She has a blind sister who has married five times, brought eleven children into the world, and been on television because at the age of sixty-three she gave birth to number twelve, a fine plump baby boy who appeared on the screen clamped onto his mother's flaccid breast. Her latest husband is twenty-two years younger than she, and for that reason this daring woman called on science and became pregnant at an age when other women are knitting for their great-grandchildren. When reporters asked her why she had done it, she replied, "So my son will be there for my husband when I die." To me that seemed very noble on her part; when I die, I hope Willie takes it very hard, and misses me a lot.

The Perverted Dwarf

We were invited to a cocktail party in San Francisco, and I went reluctantly, only because Willie asked me to. A cocktail party is a terrible trial for someone of my stature, especially in a country of tall people; it would be different in Thailand. The best idea is to avoid such events; the guests stand around, crushed together with no air, a glass in one hand and an unidentifiable hors d'oeuvre in the other. In high heels, I come up to the women's breast bone and the men's belly button; the waiters go by with their trays above my head. There is no advantage in being five feet tall, unless it's that it's easy to pick up things that fall to the floor, and in the era of the miniskirt I could make dresses from four of my first husband's neckties. While Willie, surrounded with admiring women, devoured prawns at the buffet table and told stories of his youth—such as hitchhiking around the world and sleeping in cemeteries—I dug into a corner so no one would step on me. At these events, I can't take a bite; the things I drop and spill, and those falling from other guests, fly straight to me. That evening a very amiable gentleman came toward me and when he looked down was able to make me out against the pattern of the carpet and from his Anglo-Saxon heights offer me a glass of wine. "Hello, I'm David, pleased to meet you."

"Isabel, the pleasure is mine," I replied, looking at the glass with apprehension; you can't get red wine stains out of white silk.

"What do you do?" he asked in the spirit of beginning a conversation.

That question lends itself to several responses. I could have said that right at the moment I was silently cursing my husband for having brought me to the damn party, but I opted for something less philosophical.

"I am a novelist."

"Really! How interesting! When I retire I'm going to write a novel," he told me.

"Is that right! And what is your line of work now?"

"I'm a dentist," and he handed me his card.

"Well, when I retire, I'm going to pull teeth," I replied.

Anyone could say that writing novels is like planting geraniums. I spend ten hours a day nailed to a chair, turning sentences over a thousand and one times in order to tell something in the most effective way. I suffer over the plots, I become deeply involved with the characters, I do my research, I study, correct, edit, I revise translations, and in addition travel the world promoting my books with the tenacity of a street vendor. In the car going home, driving over the superb Golden Gate Bridge, bright in the moonlight, I told Willie, laughing like a hyena, what the dentist had said, but my husband didn't see the joke.

"I'm not planning to wait till I retire. Very soon I'm going to begin writing my own novel," he announced.

"Jesus! Can you believe how arrogant some people are! And may one know what your little novel is going to be about?" I asked.

"About an oversexed dwarf."

I thought at first that my husband was beginning to catch on to my Chilean sense of humor, but he actually meant it. A few months later, Willie began to write by hand on lined yellow paper. He went around with a pad under his arm, and showed what he was writing to anyone who wanted to see it, except me. He wrote in airplanes, in the kitchen, in bed, while I teased him unmercifully. A perverted dwarf! What a brilliant idea! The irrational optimism that has served Willie so well in his life once more kept him afloat, and he was able to ignore my Chilean sarcasm, which is like those tsunamis that erase everything in their path. I thought that his literary zeal would evaporate as soon as he found how hard it is to write, but nothing stopped him. He completed an abominable novel in which a frustrated love, a legal case, and the dwarf were intertwined, confusing the reader, who couldn't determine whether he was reading a romance, a lawyer's memoir, or a string of repressed adolescent hormonal fantasies. The women friends who read it were very frank with Willie: he should take the bloody dwarf out and maybe he could save the rest of the book, if he rewrote it with more care. Male friends counseled him to take out the romance and go into more detail about the dwarf's depravation. Jason told him to take out the romance, the courts, *and* the dwarf, and write a story set in Mexico. Me? Something unexpected happened with me. The dreadful novel increased my admiration for Willie because in the process I could appreciate more than ever his basic virtues: strength and perseverance. As I have learned a few things in the years I've been writing—at least I've learned not to repeat the same errors, though I always invent new ones—I offered my husband my services as an editor. Willie accepted my comments with a humility that he doesn't have in other aspects of life, and rewrote the manuscript; it was my opinion, however, that the second version also presented too many fundamental problems. Writing is like magic tricks; it isn't enough to pull rabbits from a hat, you have to do it with elegance and in a convincing manner.

Prayers

With a grandmother like mine, who quite early instilled in me the idea that the world is magic, and that all the rest is man's delusion of greatness, given that we control almost nothing, know very little, and have only to take a quick look at history to understand the limits of the rational, it isn't strange that all things seem possible to me. Thousands of years ago, when she was still alive and I was a frightened little girl, my grandmother and her friends invited me to their spiritist sessions, surely behind my mother's back. They would place two cushions on a chair so I could reach the table-top, the same lion-footed oak table I have in my house today. Although I was very young and have no memory of it, only fantasy, I see the table jumping, moved by the spirits invoked by those ladies. The table, nonetheless, has never moved in my house; it sits in place, as heavy and categorical as a dead ox, fulfilling the modest duties of ordinary tables. Mystery is not a literary device, the salt and pepper for my books, as my enemies have accused; it is a part of life itself. Profound mysteries like the one my Sister of Disorder Jean recounted about walking barefoot over red-hot coals are transforming experiences, because they have no rational or scientific explanation. "At that moment, I knew that we have incredible capacities just as we know how to be born, to give birth, and to die," Jean said. "So, too, we know how to respond to the red-hot coals that lie in our path. After that experience, I am calm about the future. I can face the worst crises if I relax and let the spirit guide me." And that was what Jean did when her son died in her arms: she walked over fire without being burned.

Nico has asked me why I believe in miracles, spirits, and other dubious phenomena. His pragmatic mind requires proof more convincing than the anecdotes of a great-grandmother buried more than half a century ago, but to me the immensity of what I cannot explain inclines me toward magical thought. Miracles? It seems to me that they happen all the time, like the fact that our tribe keeps paddling along in the same boat, but according to your brother, they are a mixture of perception, opportunity, and a desire to believe. You, on the other hand, had my grandmother's spirituality, and you sought the answer to everyday miracles in the Catholic faith, since you were brought up in it. You were harassed by many doubts. The last thing you told me before you sank into a coma was, "I'm looking for God and I can't find him. I love you, Mamá." I want to think that you have found him, daughter, and that perhaps you were surprised since he wasn't what you expected.

Here in this world you left behind, men have kidnapped God. They have created absurd religions that have survived for centuries—I can't understand how—and continue to grow. They are implacable; they preach love, justice, and charity, and commit atrocities to impose their tenets. The illustrious gentlemen who propagate these religions judge, punish, and frown at happiness, pleasure, curiosity, and imagination. Many women of my generation have had to invent a spirituality that fits us, and if you had lived longer, maybe you would have done the same, for the patriarchal gods are definitely not suitable for us: they make us pay for the temptations and sins of men. Why are they so afraid of us? I like the idea of an inclusive and maternal divinity connected with nature, synonymous with life, an eternal process of renovation and evolution. My Goddess is an ocean and we are drops of water, but the ocean exists because of the drops of water that form it.

My friend Miki Shima practices the ancient Shinto religion of Japan, which proclaims that we are perfect creatures created by the Goddess-Mother to live in happiness; none of that guilt, punishment, penitence, hell, sin, karma; no need for sacrifices. Life is to be celebrated. A few months ago Miki went to Osaka for a Shinto ten-day training along with a hundred Japanese and five hundred Brazilians who arrived with the exuberance of Carnival. Practice began at four in the morning with chanting. When the male and female spiritual masters told the crowd gathered in that enormous, simple wood temple that each of them was perfect, the Japanese bowed and thanked them, while the Brazilians howled and danced with elation, as they would if Brazil scored a goal in the world soccer championship. Every morning at dawn Miki goes out to the garden, bows, and with a chant greets the new day and the millions of spirits that inhabit it. Then, after sushi and herbal soup for breakfast, he goes to his office, laughing in his car. Once he was stopped by a patrol car because the officers thought he was drunk. "I'm not drunk, I am doing my spiritual practice," Miki explained. The policemen thought he was mocking them. Happiness is suspicious.

Only recently we went with Lori to hear an Irish Christian theologian. Despite the obstacles of his accent and my ignorance, I did take something away from his talk, which began with a brief meditation. He asked the audience to close their eyes, relax, be aware of our breathing, in short, the usual directions in these situations, and then for us to think of our favorite place—I chose a tree trunk in your forest—and of a figure that comes to us and sits facing us. We were to sink into the bottomless gaze of that being that loved us just as we were, with our defects and virtues, without judging us. That, said the theologian, was the face of God. The person who came to me was a woman about sixty, a rotund African woman with firm flesh and a pure smile, mischievous eyes, skin as gleaming and smooth as polished mahogany, smelling of smoke and honey, a being so powerful that even the trees bowed in a sign of respect. She looked at me as I looked at you and Nico and my grandchildren when you were little, with total acceptance. You were perfect, from your transparent ears to your wet diaper odor. I wanted you to stay forever faithful to your essence, to protect you from all evil, take you by the hand and lead you until you learned to walk on your own. That love was pure happiness and celebration, although it contained the anguish of knowing that each instant that went by changed you a little and distanced you from me.

Finally we could do the tests to learn if my grandchildren carried porphyria. The Sisters of Disorder in California, and Pía and my mother in Chile, had for years been praying for my family, and I have asked myself if the prayers had done any good. The most rigorous tests have been made with ambiguous conclusions; there is no certainty that prayer has an effect. This must be a blow for the religious people who dedicate their lives to praying for the good of humankind, but it has done little to discourage either my Sisters of Disorder, or me. We do it just in case. Lucille, Lori's mother, had been diagnosed with breast cancer while I was on a book tour in the land of Christian extremism, the Deep South of the United States. At that same moment, Willie was flying the length of Latin America with a friend in a plane no bigger than a tin dragonfly, a demented trip from California to Chile.

Forty million Americans define themselves as born-again Christians, and most of them live in the center and south of the country. Minutes before my lecture, a girl came up to me and offered to pray for me. I asked her if instead of doing it for me, she would pray for Lucille, who was in the hospital that day, and for Willie, my husband, who could lose his life in some crevasse in the Andes. She took my hands, closed her eyes, and began a loud litany, attracting other people, who joined the circle, invoking Jesus, filled with faith and with the names of Lucille and Willie in every sentence. After my speech I called Lori to ask how her mother was doing and she told me that there hadn't been any operation; they examined her before taking her into the operating room and could not locate the tumor. That morning they had done eight mammograms and a sonogram. Nothing. The surgeon, who already had his gloves on, decided to postpone the intervention till the following day, and sent Lucille to another hospital where there was a scanner. They found nothing there, either. They couldn't explain it, because only days before a biopsy had confirmed the diagnosis. This would have been a verifiable miracle of prayer if only two weeks later the tumor hadn't reappeared. Lucille had her surgery anyway. However, that same day, when Willie was flying over Panama, there was a sharp change of pressure and the plane dropped two thousand meters in a few seconds' time. The skill of Willie's friend, who was piloting that big fragile insect, saved them by a hair from a spectacular death. Or was it the good thoughts of those Christians?

In spite of the prayers of my friends, and of all I asked of you, Paula, the results of the porphyria texts for Andrea and Nicole came as bad news. The condition is more serious in women than in men, since inevitable hormonal changes can trigger a crisis. We would have to live with the fear of another tragedy in the family. Nico reminded me that porphyria is not debilitating, nor does it affect normal life; the risk is increased only by certain stimuli, which can be avoided. Your case, Paula, was a combination of circumstances and error, incredibly bad luck. "We will take precautions without overdoing it," your brother said. "This is an inconvenience, but there is something positive about it: the girls will learn to take care of themselves, and it will be a good excuse for not letting them get too far away. The threat will bring us closer together." He assured me that with all the advances in medicine, the girls would have good health, children, and a long life; research in genetic engineering promises to prevent porphyria from passing to the next generation. "It's much less serious than diabetes, and other hereditary conditions," he concluded.

By then my relations with Nico had got past the reefs of previous years. We maintained the same close contact, but I had learned to respect him and honestly tried not to irritate him. My love for my three grandchildren was a true obsession, and it cost me many years to accept the fact that they weren't mine, they belong to Nico and Celia. I don't know why it took me so long to learn something obvious, something all the grandmothers in the world know without needing to be taught by a psychiatrist. Your brother and I went together to therapy for a while and even drew up written contracts establishing boundaries and rules for a peaceful coexistence, though we couldn't be too strict. Life isn't a photo in which we arrange things to their best advantage and then fix that image for posterity; it's a dirty, disorderly, quick process filled with unforeseen events. The one certainty is that everything changes. Despite our contracts, problems inevitably arose, so it was futile to worry, talk things over too thoroughly, or try to control every last detail; we had to let ourselves be carried in the flow of everyday life, counting on luck and our good hearts, because neither of us would deliberately hurt the other. If I failed—and I often did—Nico reminded me of it with his characteristic gentleness, and we didn't let it come between us. For years we've seen each other almost every day, but I am always amazed by that tall, muscular man with touches of gray hair and a peaceful air. If it hadn't been for his undeniable resemblance to his paternal grandfather, I would seriously suspect that there had been a switch in the hospital when he was born, and that somewhere there was a family with a short, explosive male who carried my genes. Nico's life improved greatly when he left the job he'd held for years. The corporation decided to outsource their work to India, where costs were much cheaper, and dismissed their employees, with the exception of Nico, who was to stay on and coordinate the programs with the office in New Delhi. He chose, however, to leave out of solidarity with his fellows. He was hired by a bank in San Francisco and also began to make transactions on the stock market, quite successfully. He has the instinct and cool head for that work, just as Lori and I had suggested some time before, but we didn't drag that up; just the opposite; we asked him where on earth he'd got such a good idea. He scorched us with a look that would shatter glass.

The Golden Dragon

The apogee of religious fanaticism gave me the theme for the second volume of the trilogy for young readers. The Christian evangelical right, which Republicans mobilized so successfully in winning presidential elections, had always been present in large numbers but had not determined this country's politics, which had a solid secular base. During the presidency of George W. Bush, the evangelicals did not achieve all the matters on their agenda, but they did effect notable changes. For example, in many educational institutions there is no mention of evolution; it has been replaced by "intelligent design," a euphemism for the biblical explanation of Creation. For them the world is ten thousand years old, and any evidence to the contrary is heresy. Guides in the Grand Canyon must be careful when they inform tourists that they can read two billion years of natural history in the geological layers. If in Norway scientists discover twenty fossils of marine animals the size of a bus and from an age prior to the dinosaurs, believers attribute that news to a conspiracy of atheists and liberals. The fundamentalists oppose abortion, in fact any form of birth control other than abstinence, but do not mobilize against war or the death penalty. Several Baptist preachers have insisted on the subjection of woman to man, erasing a century of feminist struggle. Thousands of families homeschool their children to prevent their being contaminated with secular ideas in public schools; these young people then attend Christian universities, and 70 percent of the White House interns during the Bush administration came from those universities. I hope they don't become the political leaders of the future!

My grandchildren live in the bubble of California, where such movements are a curiosity, like polygamy among some of the Utah Mormons, but they are learning because they listen to the adults in the family talking. My assignment for them was to think about an inclusive philosophy, a purified form of spirituality opposed to the extremist strain of any tendency. I didn't as yet have a clear vision of the book, but I was refining it in conversations with my grandchildren and in walks with Tabra, which during those months were almost every day because she was still going through the sorrow of having lost her father. She recalled entire poems and the names of plants and flowers he had taught her when she was a child.

"But why don't I see him the way you see Paula?" she asked.

"I don't see her, I feel her inside me. I imagine that she's always with me."

"I don't even dream about him. . . ."

We talked about the books he'd liked and others he'd been unable to teach because of censorship in the college where he worked. Books, always books. Tabra swallowed her tears and lighted up with enthusiasm when we discussed the theme of my next novel. It had occurred to her that the model for the mythic country I wanted could be Bhutan, or the Kingdom of the Thunder Dragon as its inhabitants call it, a country she had visited in her tireless pilgrimages. We changed the name to *Kingdom of the Golden Dragon,* and she suggested that the dragon could be a magical statue able to predict the future. I liked the idea that each book was set in a different continent and culture, and to imagine the place I found my inspiration in one trip we had made to India and another to Nepal, fulfilling a promise I'd made you years ago, Paula. You thought that India is a psychedelic experience, and in fact it was. The same was true of Africa and the Amazon. I'd thought that what I'd seen was so alien to my reality that I could never use it in a book, but the seeds had been germinating inside me and the fruit finally appeared in the trilogy for young readers. As Willie says: you use everything sooner or later. If I'd never been to that part of the world I couldn't have created its color, ceremonies, clothing, landscape, people, religion, or way of life. Again the help of my grandchildren was extremely valuable. We invented a religion, borrowing from Tibetan Buddhism, animism, and the books of fantasies they'd read. Andrea and Nicole go to a rather liberal Catholic school in which the search for truth, spiritual transformation, and service to others are more important than dogma. The girls landed there lacking any religious instruction. In the first week Nicole had to write a paper explaining original sin.

"I don't have any idea what that is," she said.

"I'll give you a clue, Nicole," Lori offered. "It comes from the story of Adam and Eve."

"Who are they?"

"I think sin has something to do with an apple," Andrea interrupted, without much conviction.

"Don't they say that apples are good for your health?" Nicole rejoined.

So we dismissed original sin and sat down to talk about the soul, and in that way we outlined the spirituality of the Kingdom of the Golden Dragon. The two girls were attracted to the idea of ceremonies, rituals, and tradition, and Alejandro to the possibility of developing paranormal abilities such as telepathy and telekinesis. With those leads I began writing, and every time I lost inspiration, I remembered the *ayahuasca* and my own childhood, or I went back to Tabra and the children. Andrea contributed to the plot line. Alejandro imagined the obstacles that protected the statue of the dragon: labyrinth, poisons, serpents, traps, knives, and lances that drop from the ceiling. The yetis were the creation of Nicole, who had always wanted to see one of those purported giants of the eternal snows. Tabra brought in the "blue men," a criminal sect she had heard about during a trip to the north of India.

With my outstanding team of collaborators I finished the second juvenile novel in three months, and decided that in my remaining time I would put the finishing touches on a small book about Chile. The title, *My Invented Country*, made it clear that my narrative was not a work of cool-headed scientism but, rather, of my subjective vision. With the distance of time and geography, my memories of Chile are covered with a golden patina, like the altarpieces of colonial churches. My mother, who read the first version, was afraid that the ironic tone of the book would fall flat in Chile, where in most cases the critics skin me alive. "This is a country of solemn fools," she warned me, but I knew it wouldn't be like that. Literati are one thing, but Chileans who are free of intellectual puffery are another; through the centuries we have developed a perverse sense of humor in order to survive in that land of cataclysms. In my time as a journalist, I learned that nothing is as entertaining to us Chileans as making fun of ourselves, though we would never allow it from a foreigner. I was not mistaken, because my book was published a year later without any tomato tossing. Furthermore, it was pirated. Two days after its publication, stacks of the pirated edition appeared on the streets of central Santiago—sold at a quarter of the official price—next to mountains of CDs, videos, and knockoffs of designer sunglasses and handbags. From the moral and economic point of view, pirating is a disaster for publishers and authors, but in a certain way it is also an honor; it means that there are a lot of interested readers and that the poor can buy it. Chile keeps up with progress. In Asia, the Harry Potter books are pirated so brazenly that volume seven is already on the street, a book the author has not as yet envisioned. That means that in some mysterious attic there is a little Chinese person writing as J. K. Rowling, but without the glory.

The Chile of my loves is that of your youth, when you and your brother were small, when I was still in love with your father, was working as a journalist and we lived pressed together in a little prefabricated house with a straw roof. In that period it seemed that our destiny was well set out and that nothing bad could happen to us. The country was changing. In 1970 Salvador Allende was elected president, and that occasioned a political and cultural explosion. People poured out into the street with a feeling of power they'd never had before; the young painted socialist murals, the air was filled with songs of protest. Chile was divided and families were divided as well, which is what happened in ours. Your granny marched at the head of the protests against Allende but diverted the column so it wouldn't pass by our house and throw stones. That era was, furthermore, a time of feminism and sexual revolution that affected social behavior almost more than politics; for me it was fundamental. Then came the military coup of 1973 that unleashed a wave of violence that destroyed the little world in which we felt so safe. What would our destiny have been without that coup and the years of terror that followed? What would have happened if we'd stayed in Chile during the dictatorship? We would never have lived in Venezuela, you wouldn't have met Ernesto or Nico Celia, I might not have written books or had the opportunity to fall in love with Willie, and today I would not be in California. Such musings are futile. Life goes along without a map and there is no way to turn back. *My Invented Country* is an homage to the magical land of the heart and memories, the poor cordial country where you and Nico and I spent the happiest years of childhood.

The second volume of my trilogy for young readers was already in the hands of various translators, but I couldn't concentrate on the book about Chile because I was bothered by a recurrent dream. I dreamed that a baby was trapped in a labyrinthine cellar criss-crossed with pipes and cables, like my grandfather's cellar where I spent so many hours of my childhood in solitary games. I could get *to* the baby but couldn't bring him up to the light. I told Willie about it, and he reminded me that I dream about babies only when I'm writing; surely it was something to do with my new book. I was afraid that the dream referred to the *Kingdom of the Golden Dragon*, so I reviewed the manuscript one more time, but nothing caught my eye. The dream continued to bother me for weeks, until I received the English translation and I could read it from the distance of another language; then I could see that there was a fatal problem with the plot. I had proposed that Alexander and Nadia had certain information it was impossible for them to have. And that information determined the ending. I had to ask all my translators to send back their translations, and change one chapter. Without the baby trapped in a subterranean maze, wearing my patience thin night after night, that error would have escaped me.

Disastrous Mission

The theme of the third volume of my trilogy for young readers emerged spontaneously from a peace march we attended as a family following a meeting in a famous Methodist church in San Francisco: the Glide Memorial Church. A mixture of races, ideas, and even religions come together there; it is a favorite place for Buddhists, Catholics, Jews, Protestants, and an occasional Muslim or agnostic eager to participate in a celebration of songs and embraces more than prayers. The pastor is a formidable African-American who stirs every heart with his passionate sermons on peace, a word that at that moment had antipatriotic connotations. That day, the entire congregation, on its feet, applauded until their palms were bruised, and at the end of the service many went outside to join in protesting the Iraq War.

My tribe, including Celia, Sally, and Tabra, congregated in the midst of the huge crowd milling through the streets of San Francisco. The children had painted posters, I was holding onto to Andrea to keep from losing her in the hubbub, and Nicole was sitting on her father's shoulders. It was a sunny day, and people were in a festive mood, maybe because we were cheered to see there were so many of us dissidents. However, the fifty or a hundred thousand people in the heart of San Francisco were but a flea on the back of the empire. This country is a continent divided into parcels; it is impossible to measure the magnitude or variety of reactions because each state and each social, ethnic, or religious group is a nation within a nation, all beneath the broad umbrella of the United States, "the land of the free and the home of the brave." The part about the brave seemed ironic at that moment when fear was so prevalent. Ernesto had to shave off his beard so he wouldn't be taken off the plane every time he tried to fly; anyone with the physical characteristics of an Arab was suspicious. It occurs to me that the al-Qaeda terrorists were the ones who were most surprised with the success of their strike. They planned to punch a hole in the towers; they never imagined those monumental buildings would collapse. I suppose that had it gone according to plan, the reaction would have been less hysterical and the government would have made a more realistic assessment of the enemy's strength. This was a matter of a small group of guerrillas in some distant caves, a primitive, fanatical, and desperate people who didn't have the resources to intimidate the United States.

Andrea's poster read "Words, not bombs." For a girl of ten who was beginning to write her first novel, words were undoubtedly powerful. I asked her what "Words, not bombs" meant, and she told me that her teacher had asked the class to propose ways to resolve conflict without violence. She thought about her father and herself, how as a little girl she'd had fuming fits of anger and had struck out blindly. "I have a bull inside me," she would say after her fury subsided. At those moments, Nico would gently take her arms, kneel to look into her eyes, and talk to her in a calm tone until her rage passed, a system that with some variations he always uses in critical situations. He took a course in nonviolent communication and not only does he apply what he learned to the letter of the law, he takes the refresher course every two years so he will act appropriately in an emergency. When she reached puberty, Andrea succeeded in controlling the bull, and her personality changed. "It's no fun to pester my sister anymore," Alejandro confessed when he saw he couldn't make her lose her temper. Andrea had a point. Words could be more efficient than fists. The plot of the third book would be taming the bull of war. My grandchildren and I spread out a map on my grandmother's table of the spirits to see where we would situate Alexander Cold and Nadia Santos's last adventure. The Middle East was very visible, it was what we saw every day in the news; however, the most widespread and brutal violence was taking place in Africa, where genocide is practiced with impunity. So it would be an adventure in equatorial Africa, in an isolated village where an out-of-control military man imposes terror and enslaves Pygmies. I didn't have to rack my brains to come up with the title: *Forest of the Pygmies.* Tabra, who never fails when it's time for inspiration, lent me a book of photographs of kings of African tribes, each in his fabulous robes. Most of these rulers exercised symbolic and religious, though not political, power. In some cases the king's health and fertility represented the health and fertility of the people and the land, and for that reason he was drastically disposed of the moment he became ill or old . . . unless he had the decency to commit suicide. In one tribe, the king was allowed only seven years on the throne; then he was sent to a better life and his successor ate his liver. One of the monarchs boasted that he had engendered one hundred and seventy children, and another was photographed with his harem of young wives, all pregnant; he, decked out in a lion-skin cape, feathers, and several necklaces of solid gold; the wives stark naked. There were a couple of powerful queens in the book who had their own harem of young girls, but the text did not explain who impregnated those concubines.

I did a lot of research for that book, and the more I read the less I knew and the farther away were the horizons of that enormous continent of six hundred million persons spread across forty-five countries and five hundred ethnicities. Finally I locked myself in my study and sank into the realm of magical thought. I flew directly to a swampy jungle of equatorial Africa, where miserable Pygmies, with the aid of gorillas, elephants, and spirits, were trying to rid themselves of a psychopathic king. Writing tends to be prophetic. Months after the publication of *Forest of the Pygmies*, a colonel as savage as the one in my book took over a region north of Congo, a swampy forest where he kept the Bantu population terrorized and was exterminating the Pygmies to facilitate the traffic of diamonds, gold, and arms. There was even talk of cannibalism, something I hadn't dared include in my book out of respect for my young readers.

Yemayá and Fertility

S pring of 2003 unleashed a collective reproductive
frenzy in my family. Lori and Nico, Ernesto and
Giulia, Tong and Lili, all wanted to have children, but
as by a bizarre coincidence none were able to achieve it
by traditional methods, they had to call on the discover-
ies of science and technology . . . very expensive methods that became mine to finance. They had warned me
in Brazil that I belonged to the goddess Yemayá, one of
whose virtues is fertility, and women who want to be
mothers go to her. There were so many fertility drugs,
hormones, and sperm floating in the air that I was afraid
that I myself might get pregnant. The year before, I had
secretly consulted my astrologer because I wasn't having
any dreams. I had always known from my dreams how
many children and grandchildren I was going to have,

even their names, but now, no matter how hard I tried, I had no nocturnal vision to give me a clue about those three couples. I don't know the astrologer personally, I only have her telephone number in Colorado, but I trust her because without ever having seen us she describes my family as if it were hers. The only person whose astral chart she hasn't done is Nico's, and that because I don't remember what hour he was born and he refuses to let me have his birth certificate. The woman told me that this son was my best friend and that we had been married in a previous incarnation. Understandably, Nico didn't want to hear of such a horrendous possibility, and that's why he hides the certificate. Your brother doesn't believe in reincarnation because it is mathematically impossible, or astrology, of course, but he thinks it reasonable to take precautions, just in case. . . . I don't believe every last bit of it, either, but there's no reason to block out such a useful tool for literature.

"How do you explain that the woman knows so much about me?" I asked Nico.

"She looked you up on the Internet, or she read *Paula*."

"If she researched every client in order to fake them out, she would need a team of assistants and would have to charge a lot more. No one knows Willie, he's not on the Internet, but she was able to describe him physically. She said he was tall, with broad shoulders, a large neck, and handsome."

"That's very subjective."

"How can it be subjective, Nico! No one would say my brother Juan is tall, has broad shoulders, a thick neck, and is handsome."

In the end, I get nowhere by discussing such subjects with my son. The fact is that the astrologer had already told me that Lori could not have children of her own but that "she would be the mother of several children." I interpreted that to mean that she would be the mother of my grandchildren, but apparently there were other possibilities. About Ernesto and Giulia she said that they should not make the attempt until spring of the following year, when the stars were in the ideal position; any earlier would have no result. Tong and Lili, on the other hand, would have to wait a lot longer, and it was not certain that the baby would be theirs, it might be adopted. Ernesto and Giulia decided to obey the stars and wait until spring to begin the fertility treatments. Five months later, Giulia was pregnant; she swelled up like a dirigible, and soon learned she was expecting twin girls.

One day Juliette, Lori, Giulia, and I were in a restaurant, and Lori was telling how about half the young women she knew, including her hairstylist and her yoga teacher, were either pregnant or had just had a baby.

"Do you remember when I said I would have a baby for you, Isabel?" Juliette asked.

"Yes. And I told you that I would be crazy to have a child at my age."

"That time I said I would have it only for you, but I think now that I would also do it for Lori."

A moment of absolute silence fell over the table as Juliette's words made their way to Lori's heart, who burst into tears when she realized what her friend had just offered. I don't know what the waiter thought, but on his own he brought us chocolate cake, courtesy of the house.

Then began a complicated process that Lori, with her extraordinary perseverance and organization, directed for nearly a year. First it had to be decided whether or not Nico would be the father, because of the risk of porphyria. After talking it over between them, and with the family, they agreed that they were will-ing to take the chance; it was important to Lori that the baby be Nico's child. Then they had to obtain an egg. It couldn't be Juliette's because if she knew she was the biological mother she would not be capable of giving up the baby. Through the Internet they chose a young Brazilian donor who had a slight resemblance to you, Paula, a family look. She and Juliette had to undergo large doses of hormones, the former to ensure several eggs to be harvested, and the latter to prepare her womb. The eggs were fertilized in a laboratory, then the embryos were implanted in Juliette. I feared for Lori, who might suffer yet another frustration, and especially for Juliette, who now was over forty and a widow with two growing boys. If something happened to her, what would become of Aristotelis and Achilleas? As if she had read my mind, Juliette asked Willie and me to look after her children should any misfortune befall her. We had reached the boundaries of magical realism.

Traffic in Organs

Lili, Tong's young wife, endured her mother-in-law's abuse for a year, until her submissiveness came to an end. If her husband hadn't intervened maybe Lili would have strangled that lady with her bare hands—an easy enough crime because the woman had a neck like a chicken. There must have been quite an uproar; the San Francisco Police Department sent one officer who spoke Chinese to separate the family members at that address. By then Lili had demonstrated that she had been serious when she said she hadn't come to America for the visa but to form a family. She had no intention of getting a divorce, despite the mother-in-law and the habitual unpleasantness of Tong, who still was suspicious that she would ask for a divorce as soon as the period stipulated by law had passed.

Tong realized that the submissive wife he had ordered by mail was a strapping female warrior. His mother, frightened for the first time in her seventy-plus years, said that she could not keep living with that daughter-in-law who at the first moment she dropped her guard might send her to meet her ancestors. She forced Tong to choose between his wife, that brutish woman obtained through "questionable electronic channels," as she said, and her, his legitimate mother, whom he had lived with all his life. Lili did not give her husband long to think about it. She stood firm, and it was not she who left the house but her mother-in-law. Tong installed his mother in an apartment for seniors in the center of Chinatown, where she plays mah-jong with other ladies her age. They sold the house and bought a smaller, more modern one near where we live. Lili rolled up her sleeves and threw herself into the task of converting it into the home she had always wanted. She painted the walls, pulled the weeds in the garden, decorated with white, starched curtains, simple, well-made furniture, plants, and fresh flowers. And with her own hands she installed bamboo floors and French windows.

I learned these details little by little, through sign language, drawings, and the few English words that Lili and I shared, until summer came and my mother arrived from Chile and in less than five minutes was sitting in the living room with Lili, having tea and talking like old friends. I don't know what language they were using; Lili doesn't speak Spanish, my mother doesn't know Mandarin, and the English of both leaves something to be desired.

Two days later, my mother told me that we were invited to have dinner at Lili and Tong's house. I explained that that was impossible, she must have misunderstood. Tong had worked half a lifetime with Willie, and the only social event he had ever shared with us was Nico's wedding, and that because Lori had forced him to come. "That may be, but tonight we're having dinner with them," my mother replied. She was so insistent that to appease her we went. I was thinking that we would ring the doorbell with some excuse and she would find she'd been mistaken, but when we got there we saw Lili sitting outdoors, waiting for us. Her house was dressed for a party with bouquets of flowers, and in the kitchen were a dozen different dishes she was putting the last touches to. Her chopsticks were flying through the air, transferring ingredients from one pot to another with magical precision, while my mother, installed in the place of honor, chatted with her in their Martian tongue. A half hour later, Willie and Tong arrived, and for the first time I had an interpreter and was able to communicate with Lili. After we had devoured the banquet, I asked her why she had left her country, her family, and her job as a surgical nurse to run the risky adventure of blindly marrying and moving to America, where she would always be a foreigner.

"It was because of the executions," Tong translated.

I assumed there had been a linguistic error—after all, Tong's English isn't much better than mine—but Lili repeated what she'd said and then, with the help of her husband and a lot of miming, she explained why she had joined the thousands of women who leave their country to marry a stranger. She told us that every three or four months, when the prison notified them, she had to accompany the chief surgeon of the hospital to the executions. They left in his car, carrying a large box filled with ice, and traveled four hours along rural roads. At the prison they were led to a basement where half a dozen prisoners were lined up waiting, hands tied behind their backs and eyes blindfolded. The commander gave an order and the guards shot the prisoners in the temple at point-blank range. The minute the bodies fell, the surgeon, helped by Lili, hastily tore out organs for transplant: eyes to provide corneas, kidneys, livers, in short, anything that could be used. They returned from that butchery covered with blood, and with the ice chest filled with organs that then disappeared on the black market. It was a prosperous business, organized by certain physicians and the director of the prison.

Lili told that macabre story with the eloquence of a consummate actress of the silent screen: she rolled her eyes back, shot herself in the head, fell to the floor, picked up a scalpel, cut and ripped out organs, everything in such detail that my mother and I were overcome by an attack of nervous giggles, to the horror of the others, who didn't understand what the devil we found so comical. Our laughter reached the level of hysteria when Lili added that on the last trip the car turned over as they were returning from the prison and the surgeon had died instantly, leaving Lili abandoned in open country with a cadaver clutching the wheel and a load of human organs resting in ice. I have often wondered whether we truly understood Lili's tale, whether it was her idea of a joke or if in fact that enchanting woman who picks my grandchildren up from school and looks after my dog as if it were her daughter actually went through those hair-raising experiences.

"Of course it's true," was Tabra's opinion when I told her. "In China there is a concentration camp associated with a hospital where thousands of people have disappeared. They rip out the organs while the 'donors' are still living, then cremate the bodies. The refugees who work in my studio have equally terrifying stories. In their countries, poor people sell their kidneys to feed their children."

"And who buys them, Tabra?"

"The wealthy, including here in America. If one of your grandchildren needed an organ to keep from dying and someone offered you one, wouldn't you buy it and not ask questions?"

That was only one of the conundrums Tabra posed during our walks in the woods. Instead of enjoying the aroma of the pines and the song of the birds, I would come home from those walks completely undone. But we didn't always talk about the atrocities committed by our fellow humans, or about politics. we also talked about Plumed Lizard, who made sporadic appearances in my friend's life only to vanish again for months. Tabra's ideal would be to have her Indian, complete with pigtails and necklaces, living in a Comanche tepee on her patio.

"That doesn't seem practical, Tabra. Who would be in charge of feeding him and washing his undershorts? He would have to use your bathroom, and then you'd have to clean up after him," I told her, but she was impervious to this kind of mean-spirited reasoning.

Children that Didn't Come

Three times they implanted in Juliette the laboratory embryos conceived from the eggs of the beautiful Brazilian donor and Nico's sperm. On all three occasions our tribe awaited the results for weeks, with souls hanging by a thread. We invoked the usual sources of magic. In Chile my friend Pía and my mother went to our national saint, Padre Hurtado, and left donations for his charitable works. The image of that revolutionary saint, which all Chileans carry in our hearts, is that of a young and energetic, black cassock-clad man with a shovel in his hand, hard at work. His smile is not in the least beatific but, rather, clearly defiant. It was he who coined your favorite phrase: Give till it hurts. Following the failure of the first two, the third embryo implant took place in the summer. Lori and Nico had for a year been planning a trip to Japan, and they decided to go anyway. If their dream of having a baby came true, it would be their last vacation for a while. They would receive the news there; if it was positive, they would celebrate, if not, they would have a couple of weeks to themselves, intimate, quiet time in which to resign themselves, far from the condolences of friends and relatives.

One of those early mornings I woke with a start. The room was palely illuminated by the subtle splendor of dawn and the night light we always leave on in the hall. No air was moving and the house was wrapped in an abnormal silence; I couldn't hear the rhythmic snoring of Willie and Olivia, or the usual murmuring of the patio palm trees dancing in the breeze. Beside my bed were two pale children, standing hand in hand, a girl about ten and a boy a little younger. They were wearing clothes from the 1900s, lace collars and patent leather high-top shoes. It seemed to me that there was sadness in their large dark eyes. We looked at each other for a second or two, but when I turned on the light they disappeared. I waited a moment, hoping in vain that they would come back, but finally, when the galloping of my heart slowed, I went on my tiptoes to call Pía. In Chile it was five hours later but my friend was still in bed, embroidering one of her patchwork bags.

"Do you think those children have anything to do with Lori and Nico?" I asked.

"No! Of course not! They're the children of the two English ladies," she replied with calm conviction.

"What English ladies?"

"The ones who visit me. The ones who walk through the walls. Haven't I told you about them?"

On the scheduled day, Lori was to call the nurse who coordinated the treatment in the fertility clinic, a woman with the vocation of a godmother, who handles each case with delicacy; she knows how much hangs in the balance for these couples. Because of the time difference between Tokyo and California, Lori and Nico set the alarm for five in the morning. As they couldn't make international calls from the room, they hurriedly dressed and went down to the front desk of the hotel, where at that moment they found no one to help them. Fortunately, Lori knew there was a telephone booth outside. They went out to a side street that during the day was seething with activity, thanks to popular restaurants and shops for tourists, but at that hour was deserted. The antiquated booth was straight out of a '50s film, and the phone could be operated only with coins, but Lori had thought ahead and brought enough change with her to call the clinic. Blood was pounding in her temples and she was trembling as she dialed the number with a prayer on her lips. Her future was being determined in those instants. From the other side of the planet came the voice of the godmother. "It didn't take, Lori. I'm so very sorry. I don't know what happened, the embryos were the very best . . . ," she said, but Lori heard no more. Stunned, she hung up the receiver, turned, and fell into her husband's arms. And that man, who at first was so resistant to the idea of bringing more children into the world, sobbed openly; he had been as passionate as she about the idea of their having a child together. They embraced without a word, and minutes later stumbled out onto the empty, silent street, gray in the predawn. Columns of steam rose from the grates in the sidewalks, lending a phantasmagoric air to the scene, a perfect metaphor for the desolation they felt. The rest of their time in Japan was spent convalescing. They had never been so close. In their shared sorrow they came together at a very deep level, naked, defenseless.

Something in Lori changed after that experience, as if a glass had broken inside her and the obsessive desire that had been her hope and her torment had drained away like water. She realized that she couldn't live with Nico if she were in a swamp of frustration. It wouldn't be fair to him. Nico deserved the kind of happy devotion he had tried so hard to build between them. She realized that she had come to the end of a tortuous road, and that she must root out her obsession about being a mother, if she was to go on living. After having tried every possible resource, it was obvious that a child of her own was not to be her destiny, but her husband's children, who had been with her for several years and who loved her a lot, could fill that void. That resignation didn't happen overnight; she was sick in body and soul for nearly a year. Lori had always been slim, but within a few weeks' time she lost so much weight that she was nothing but skin and bones, with large, sunken eyes. She injured a disk in her back and for months was close to being an invalid, trying to function with painkillers so strong that they made her hallucinate. At moments she despaired, but the day came when she emerged from that long grieving, her back healed, her soul at peace, transformed into a different woman. We all could see the change. She gained weight, looked younger, let her hair grow, painted her lips, resumed her yoga and long walks through the hills, but now as a sport, not an escape. We heard her laugh again, the contagious laugh that had seduced Nico, something we hadn't heard for a long, long time. At last she was ready to give herself to the children with all her heart, with joy; it was as if a fog had dissipated and she could see them clearly. They were hers. Her three children. The children the shells in Bahía and the astrologer in Colorado had predicted for her.

Striptease

Willie and Lori had worked together in the Sausalito brothel for years, even sharing one bathroom. It's amusing to watch the relationship between those two people who could not be more different. To Willie's chaos, cursing, and rushing about, Lori opposes order, precision, calm, and gentility. At noon, as Lori tosses her macrobiotic salad with tofu, Willie perfumes the atmosphere with the garlic of spicy sausages that would perforate the intestines of a rhinoceros. After he's taken the dog for a walk, he comes into the office wearing the muddy boots of a ditch digger, and Lori amiably cleans the stairs so some client won't slip and break his neck. Willie piles mountains of papers on his desk, from legal documents to used paper napkins, and every so often Lori sweeps through and throws them into the wastebasket; he doesn't even notice, or maybe he does notice but doesn't kick about it. They share the vice of photography and travel. They consult on everything and celebrate each other with no perceptible signs of sentimentalism; she always efficient and tranquil, he always hurrying and grumbling. She takes care of the computer, and keeps the Web page up to date, and he cooks meatballs for her following her grandmother's recipe; he shares with her everything he buys wholesale, from toilet paper to papayas, and loves her more than anyone in the family, except me . . . maybe.

Willie teases her, of course, but he also tolerates her jokes. Once Lori made up an exquisitely lettered bumper sticker she stuck on the back of his car. It read: I LOOK VERY MACHO BUT I WEAR WOMEN'S PANTIES. Willie drove around for a couple of weeks wondering why so many men were waving from other cars. Considering that we live in a part of the world that may have the highest number of homosexuals per capita, it wasn't difficult to explain. When Willie discovered the sign, he nearly had a stroke.

From time to time the alarm in the brothel goes off without any provocation, which tends to cause difficulties. Like once when Willie got there in time to hear the deafening clamor of the alarm and ran inside through the kitchen—on the lower floor—to turn it off. It was a winter evening and near dark. At the same moment he ran in, a policeman, who had kicked in the main door, came running down the steps, still wearing his sunglasses and carrying a pistol in his hand. He yelled at the top of his lungs for Willie to put up his hands. "Take it easy, man, I'm the owner," my husband tried to explain, but the cop ordered him to shut up. He was young and inexperienced and perceptibly nervous; he kept yelling and calling for backup over his phone, while the white-haired man with his face plastered against the wall boiled with rage. The incident dissolved without consequences when other armed police arrived in combat gear and, after patting Willie down, listened to what he was saying. That episode set off an endless string of curses from Willie as Lori doubled over with laughter—though she might have laughed a little less had she been the victim. A week later when we were all at work, some of Lori's friends, who were also good friends of ours, began to filter in. I thought it was a little strange, but I was on the phone with a journalist in Greece and merely waved at them from a distance. I finished my conversation just as a policeman came in—tall, young, blond, very handsome, sunglasses, and pistol at his waist—who asked to speak with Mr. Gordon. Lori called Willie, and he came down from the second floor ready to tell that "uniform" that if they kept fucking around and bothering him he was going to sue the police department. All Lori's friends stationed themselves on the stairs to watch the show.

The handsome policeman held up a bundle of papers and told Willie to have a seat because he needed to sign some forms. Grousing, my husband obeyed. Then we heard strains of Arabic music, and the man began to dance like an enormous odalisque. First he took off his hat, then his boots, then the pistol, jacket, and pants, to the absolute horror of Willie, who pushed his chair back, red as a lobster, sure than the man had escaped from some institution. The howls coming from the stairway gave him the clue that the "policeman" was an actor Lori had hired, but by then the dancer had nothing on but his sunglasses and a thong that came up short in covering his private parts.

Considering that we all work at the same site, that we run Willie's office, the foundation, and my office among us, that we see each other nearly every day, that we go together on vacations in the far corners of the planet and live within a radius of six blocks, it's surprising we get along so well. A miracle, I'd say. Therapy, is Nico's explanation.

My Favorite Writer

C ontrary to what I might have expected, my cutting comments on Willie's novel and his perverted dwarf did not provoke a war between us, which would have been the case had Willie been reckless enough to offer a negative criticism of one of my books; it was evident, however, that I wasn't the right person to help him, he needed a professional editor. About that time a young literary agent appeared who was at first very interested in the book and devoted herself to boosting my husband's ego, but little by little her enthusiasm cooled. At the end of six months, she congratulated him on the effort, assured him that he had talent, and reminded him that many authors, including Shakespeare, had written pages whose final destination was a trunk. We had several trunks in our home where the dwarf could sleep the sleep of the just indefinitely, while Willie was thinking about another subject, but he ignored everyone's opinion and sent the book to various agents who returned it with courteous, but unmistakable, rejections. Far from deflating him, those letters of condemnation reinforced his fighting spirit—my husband is not a person to be stymied by reality. This time I didn't make fun of his writing; it had occurred to me that literature could give meaning to the later years of his life. If what the agent had told him was right, and Willie had talent, and if he took the matter seriously and was capable of becoming a writer already past the age of sixty, then down the road I wouldn't have to look after a gaga old man. It would work out well for both of us. A creative life would keep him happy and healthy well into an advanced old age.

One night, embraced in bed, I explained the advantages of writing about something you know. What did he know about sodomite dwarfs? Nothing, unless he was projecting onto that lamentable character some aspect of his own personality that I didn't know about. On the other hand, he had been a lawyer for more than thirty years, and he had a formidable memory for details. Why didn't he explore the genre of the detective novel? Any of the many cases he had tried could serve as a point of departure. Nothing is as entertaining as a good, bloody murder. He lay there thinking, without saying a word. The next day we were driving through Chinatown in San Francisco and Willie saw an albino Chinese man standing on a corner. "I know what my next novel will be about. It will be a criminal case with a Chinese albino like him," he told me in the same tone he'd used at the sadomasochist fair in San Francisco where he'd seen the dwarf wearing a dog collar and first mentioned his literary aspirations. Two years later, his novel was published in Spain under the title *Duelo en Chinatown*, and several other editors bought it to translate into their various languages. We went together to the launch of his book in Madrid and Barcelona, accompanied by his sons and a couple of faithful friends eager to applaud him. Everywhere he went, the press welcomed him with curiosity, and, after talking with him, published cordial articles because he won everyone over, especially women, with his simplicity. No pretension, only the blue gaze and the dashing smile beneath the brim of his eternal hat. The day of the launch in Madrid, one of the journalists asked him if he wanted to be famous, and he answered, with deep feeling, that he already had more than he'd ever dreamed: the fact that journalists were there, and people wanted to read his book, was a gift. He disarmed them, while his publisher twisted in his chair because he'd never had such an honest author.

For once, it was my turn to carry the suitcases and repay Willie in some small part for the irksome tasks he'd performed all those years of traveling around the world with me.

"Cherish this moment, Willie, because it will never happen again. The joy of seeing the first copy of your first book is unique. If you have other publications in the future, they won't compare with this one," I warned him, remembering what I'd felt when I saw the horrendous first edition of *The House of the Spirits*, which I keep wrapped in silk paper, signed by the actors in the film and those in the London play.

Willie's barrio Spanish splashed with Mexican idioms and words in English won him points, and the rest was accomplished by his Italian Borsalino fedora, which gives him the air of a detective of the 1940s. He appeared in many magazines and newspapers, he was interviewed on several radio stations, and we have pictures of a bookstore in Spain, and another in Chile, where *Duelo en Chinatown* is displayed in the window among the best sellers. On a radio program, he mentioned the pathetic dwarf from the unpublished book, and later in the hotel a man came up to him to tell him he'd heard his interview on the radio.

"But how did you know it was me?" Willie asked, amazed.

"The interviewer mentioned your hat. I want to tell you that I have a friend who's a dwarf, and he is as perverted as the one in your novel. Pay no attention to your wife, just publish it. It will sell like hotcakes, everyone likes depraved dwarfs."

A month later, in Mexico, someone told Willie that in the 1900s there was a bordello in Juarez staffed by two hundred female dwarfs. Two hundred! He even gave Willie a book about that Fellini whorehouse. I worry that it might inspire Willie to retrieve his abominable little man from the trunk.

I had never seen Willie so happy. I will definitely not have to take care of a drooling old man, because on the plane he pulled out his pad of yellow paper and began to write another book. The same astrologer who once told Willie that his children were his worst enemies, also told him that the last twenty-seven years of his life would be very creative, so I can relax until he's ninety-six.

"Do you believe in those things?" I asked Carmen Balcells, my agent, when I told her the story.

"If you can believe in God, I don't see why you can't believe in astrology," was her answer.

A Bourgeois Couple

In February, 2004, the mayor of San Francisco had committed the unpardonable political sin of trying to legalize the union of homosexuals; that galvanized the Christian right to defend "family values." Preventing gay marriage became the rallying cry of the Republicans during Bush's reelection campaign that same year. It's astounding how that issue weighed more heavily in the voting booth than the war in Iraq. The country wasn't mature enough for an initiative like the mayor's. He had issued it on a weekend, when the courts were closed, so no judge was able to block the order. The minute the news was announced, hundreds of couples lined up at the city hall, an unending queue in the rain. During the next hours, messages of congratulations poured in and bouquets of flowers carpeted the street. The first couple to be married was two women of eighty-some years, white-haired feminists who had lived together for more than fifty years, and the second was two men, each of whom was carrying a baby in a sling on his chest: adopted twins. The people in that long line wanted a normal life, to raise children, buy a house, inherit from a mate, and be together at the hour of death. No family values there, obviously. Celia and Sally were not part of that throng; they thought that the mayor's initiative would quickly be declared illegal, which is in fact what happened.

By then it had been a long time since Celia's brother had left the scene. The strategy of marrying Sally to obtain a U.S. visa had not been effected and he had instead decided to return to Venezuela, where finally he married a pretty young girl, bossy and entertaining, had an enchanting little boy, and found the destiny that had eluded him in the United States. That had allowed Sally and Celia to be legally joined in a "domestic part- nership." I imagine that it would have been a little complicated to explain to the clerks if Sally had "mar- ried" two people with the same surname but different sexes. As for the children, who had seen the wedding photo of her and their uncle, not much explanation was required; they understood from the beginning that Sally was only doing him a favor. I think that nothing regarding the family shocks my grandchildren now.

Celia and Sally have become so stable and bourgeois that it's difficult to recognize them as the daring young women who years before defied society to love each other. They like to go to restaurants, or lie in bed watching their favorite television programs, or organize parties in their tiny home, where they somehow manage to greet a hundred people with food, music, and dancing. One of them is a night owl and the other goes to bed at eight, so their schedules don't coincide.

"We have to make a date to meet at noon, agendas in hand, or we'd live like friends rather than lovers. Finding moments for intimacy when you have three kids and so much work is a real challenge," Celia confessed, laughing.

"That's more information than I need, Celia."

They remodeled their house, converting the garage into a TV room and a room for Alejandro, who's at an age when he wants privacy. They have a dog named Poncho, black, easy-going, huge, like the Barrabás of my first novel, who sleeps on the children's beds by turns, one night with each one. His arrival terrified the two spitfire cats, which fled across the rooftops and were never seen again. When my grandchildren spend the week in their father's house, an unhappy Poncho throws himself at the foot of the stairs, with soulful eyes awaiting the following Monday.

Celia discovered the passion of her life: mountain biking. Although she's over forty, she wins prizes in endurance races competing with twenty-year-olds, and she started a small side business leading biking tours: Mountain Biking Marin. Fanatics come from remote places to follow her up rugged trails to the heights.

It's my opinion that these are two happy women. They work for a living but don't kill themselves just to make money, and they agree that their priority is the children, at least until they're grown and independent. I remember the days when Celia used to run, hide, and throw up because she was trapped in a life that wasn't right for her. They had the good luck to be living in California, at the dawn of the twenty-first century. In a different place and a different time they would have been condemned by implacable prejudices. Here their being gay does not pose a problem even at the girls' Catholic school; that's not what defines them. Most of their friends are couples, with children, ordinary, everyday families. Sally unhesitatingly took on the role of housewife, while Celia tends to behave like a caricature of a Latin American husband.

"How do you put up with her, Sally?" I asked once, when I saw her cooking and helping Nicole with her arithmetic lesson while Celia, in her scandalous pants and crazy helmet, went pedaling around mountain trails with tourists.

"Because we have such a good time together," she replied, stirring the pot.

In this adventure of forming a couple, chance plays a large role, but so does intent. Often during an interview someone will ask me "the secret" of Willie's and my notable relationship. I don't know what to answer because I don't know the formula, if indeed there is one, but I always remember something I learned from a composer and his wife who visited us. They were in their sixties but they looked young, strong, and filled with enthusiasm. This musician explained that they had married—or, more accurately, renewed their commitment—seven times during their long life together. They had met when they were university students, fallen in love at first sight, and had been together for more than four decades. They had passed through various stages, and in each one they had changed and been near the point of separating but had opted to review their relationship. Following each crisis, they had decided to stay married a little longer, for they discovered that they still loved each other even though they were not the same persons they'd been before. "In all, we have gone through seven marriages and no doubt there are more to come. It isn't the same thing to be a couple when you are raising children, with no money and no time, as when you are in your mature years, established in your profession, and expecting your first grandchild," he said. He told us, as an example, that in the 1970s, at the height of the hippie madness, they had lived in a commune with twenty idle young people; he was the only one who was working; the others spent the day in a cloud of marijuana smoke, playing the guitar, and reciting in Sanskrit. One day he grew tired of supporting them and kicked them out of the house. That had been a crucial moment when he, with his wife, had had to revise the rules of the game. Then came the materialistic stage of the 1980s, which nearly destroyed their love because they were both running after success. On that occasion, too, they had opted to make basic adjustments and start over again. And so it went, again and again. It seems to me that theirs is a formula that's right on the mark, and one Willie and I have had to put into practice more than once.

The Twins and Gold Coins

Ernesto and Giulia's twin girls were born on a splendid morning in June of 2005.

I got to the hospital at the moment that Ernesto had just welcomed his two daughters and was sitting with two rosy packages in his arms, crying. I started crying too, happy tears because these infants represented a definitive end to his being a widower and the beginning of a new stage in his life. Now he was a father. When Willie saw them, he said that one looked like Mussolini and the other like Frida Kahlo, but a couple of weeks later, their features took shape and we could see that they were a pair of beautiful little girls: Cristina, blond and happy, like her mother; Elisa, dark and intense, like her father. They are so different in looks and personality that they could have been adopted: one in Kansas and the other in Tenerife. Giulia gave herself completely to her daughters, to the degree that for more than a year she hasn't talked about anything else. She managed to get them on a schedule of sleeping and eating at the same time; that gave her a few minutes of freedom between naps that she uses to restore order from chaos. She's bringing them up with Latin music, the Spanish language, and with no fear of germs or accidents. Pacifiers go from the floor into the mouth and no one makes a fuss, and later, before they learned to walk, the twins discovered how to go up and down tiled stairs with the sharp edges of the steps scraping their bellies. Cristina is a little weasel who can't stop moving for a minute; she approaches the abyss of balconies with suicidal indifference. Elisa, on the other hand, sinks into somber thoughts that tend to bring on attacks of inconsolable tears. I don't know how, but Giulia finds the energy to dress them like dolls, in embroidered booties and sailor hats.

The previous year, precisely on December 6, the anniversary of your death, Ernesto was accepted at the university to study for his master's degree at night; at the same time he was hired to teach mathematics in the best public school in the county, fifteen minutes from his house. He had been without a job for a few months, and had gone around with a dark cloud over his head, meditating on his future. Giulia, always sparkly and optimistic, was the only one in the family who never doubted that her husband would find his way, though the rest of the family was getting a little nervous. Tío Ramón reminded me in a letter that men suffer a crisis of identity somewhere around the age of forty; it's part of the process of maturing. It had happened to him in 1945 when he fell in love with my mother in Peru, over sixty years ago. At that time he'd gone to a hotel in the mountains, locked himself in the silence of a room for days, and when he came out he was a different person. He had shaken off his Catholic religion, his family, and the woman who was then his wife. There he had grown up, been educated, had matured, and had realized that until that moment he had lived confined in the strait-jacket of social conventions. When he tore that off, he lost all fear of the future. It was he who coined the phrase he taught me in my early adolescence, and that I have never forgotten: *Everyone else is more afraid than you are.* I repeat those words when I have to face some frightening situation, from an auditorium filled with people to loneliness. I have no doubt that Tío Ramón determined his fate in that drastic way because I have seen his decisiveness on other occasions. Like the time he caught my brother Pancho smoking—he was only about ten. Tío Ramón crushed out his own cigarette butt before us and announced, "This is the last cigarette of *my* life, and if I catch any of you smoking before you're adults you will have to deal with me." He never smoked again. Fortunately Ernesto overcame his forty-year crisis, and when his daughters were born he was ready to welcome them, already settled in his position as a high school math teacher and studying to be a college professor.

Alfredo López Lagarto-Emplumado appeared on the Spanish TV channel, handsomer than ever, dressed in black with an Indian band around his forehead and several silver and turquoise necklaces around his neck. Tabra called me at ten o'clock at night to tell me to turn on the television and watch, and I had to admit that the man was very attractive, and that if I hadn't known him so well, his image on the screen would undoubtedly have impressed me. He was speaking English—with subtitles—with the calm of an academic and the moral conviction of an apostle, explaining the justice of why he had been drawn to the mission of recovering Moctezuma's crown, a symbol of the dignity and tradition of the Aztec people that had been appropriated by European imperialism. After so many years of not being heard, at last his message had reached the ears of the Aztecs and their hearts had ignited like gunpowder. The president of Mexico had sent a commission of jurists to Vienna to negotiate with the congress of that country for the return of the historic trophy. Lagarto ended by calling on Mexican immigrants in the United States to join in the struggle with their brothers and enlist the aid of the North American government to put pressure on the Austrians. I congratulated Tabra for her friend's leap to fame, but she answered with a deep sigh. If Lagarto had been elusive before, now it would be impossible to hunt him down. "Maybe he will follow me to Costa Rica after he gets the crown back," she suggested, but without conviction. "Well, that's if I ever save enough money to move there." Be careful what you ask for, I thought, heaven might grant it, but I didn't say it to her. For a long time, Tabra had been buying gold coins that she hid in nooks and crannies, creating the danger of having them stolen.

Doña Inés and Zorro

While Tabra was preparing to emigrate, I was deep into researching a subject I'd been thinking about for four years: the legendary epic of the one hundred and ten heroic rogues who conquered Chile in 1540. With them was one Spanish woman, Inés Suárez, a seamstress from Plasencia, a city in Extremadura, who traveled to the Indies following the footsteps of her husband and thus ended up in Peru, where she discovered that she was a widow. Instead of returning to Spain, she stayed in the New World and later fell in love with Don Pedro de Valdivia, an hidalgo whose dream was "earn fame and leave glory to my name," as he reported in his letters to the king of Spain. For love, and not greed for gold or glory, Inés went with him. For years I had carried in my mind the image of that woman who crossed the desert of Atacama, the most arid in the world, fought like a brave soldier against the Mapuche Indians, the most ferocious warriors in the Americas, founded cities, and died an old woman, loving another conquistador. She lived in cruel times and committed more than one brutal act, but compared to any of her companions on that adventure, she appears as honest and upright

I have often been asked where I find the inspiration for my books. I wouldn't know how to answer that. As I travel through life, I gather experiences that lie imprinted on the deepest strata of memory, and there they ferment, are transformed, and sometimes rise to the surface and sprout like strange plants from other worlds. What is the fertile humus of the subconscious composed of? Why are certain images converted into recurrent themes in nightmares or writing? I have explored many genres and diverse themes, and it seems to me that in each book I invent everything anew, including the style, but I have been doing this for more than twenty years and I can see repetitions. In nearly all my books there are defiant women, born poor or vulnerable, destined to be subjected, but they rebel, ready to pay the price of freedom at any cost. Inés Suárez is one of them. My female protagonists are always passionate in their loves and loyal to other women. They are not moved by ambition but by love; they throw themselves into adventure without measuring the risks or looking back, because to remain paralyzed in the place society holds for them is much worse. That may be why I am not interested in queens or women who come into the world in a cradle of gold, nor do I favor beautiful women who have their path paved by men's desire. You always laughed at me, Paula, because pretty women in my books die before page sixty. You said it was pure envy on my part, and I'm sure you were partly right since I would have liked to be one of those beauties who get what they want without any effort; for my novels, however, I prefer heroines with courage for whom no one does any favors; they make their way on their own. It isn't strange, therefore, that when I read about Inés Suárez between the lines in a history book—only rarely are there more than a couple of lines when women are mentioned—she piqued my curiosity. She was the kind of character that normally I have to invent. When I did my research I learned that nothing I could imagine could surpass the reality of her life. What little is know about her is spectacular, nearly magical. Soon I would begin to tell her story, but my plans were altered by three unexpected visitors.

One Saturday about noon, three people came to our home; at first we took them to be Mormon missionaries.

They weren't, fortunately. They explained that they owned the world rights to Zorro, the California hero we all know. I grew up with Zorro because Tío Ramón was an ardent admirer of his. You remember, Paula, that in 1970 Salvador Allende appointed your grandfather to serve as ambassador to Argentina, one of the most difficult diplomatic missions at that time, and he performed that duty with honor until the day of the military coup, when he resigned his post because he was not disposed to represent a tyrannical government. You often visited them. You were only seven years old but you made that plane trip by yourself. In that enormous building with its innumerable reception rooms, twenty-three bathrooms, three grand pianos, and army of employees, you felt like a princess; your grandfather had convinced you that it was his palace, and that he was royalty. During those three years of intense service in Buenos Aires, at four in the afternoon the honorable ambassador excused himself from any appointment in order to enjoy in secret the half hour of the serial version of *Zorro* on television. With that background, I could do no less than welcome those three visitors with open arms.

Zorro was created in 1919 by Johnston McCulley, a California author of dime novels, and since then has lived in the popular imagination. *The Curse of*

Capistrano narrated the adventures of a young Spanish hidalgo in Los Angeles in the nineteenth century. By day, Don Diego de la Vega was a hypochondriacal and frivolous young man, but by night he dressed in black, donned a mask, and was converted into Zorro, avenger of Indians and the poor.

"We've seen Zorro in every form: films, television serials, comic books, costumes, everything except a literary work. Would you like to write that?" they proposed.

"What are you talking about! I'm a serious writer, I don't write on commission," was my first reaction.

But then I remembered Tío Ramón, and my honorary grandson Achilleas disguised as Zorro for Halloween, and the idea began to haunt me, so much so that Inés Suárez and the conquest of Chile had to wait their turn. According to the owners of the rights to Zorro, the project fit me like a glove: I'm Hispanic, I write in Spanish, I know California, and I have some experience of writing historical and adventure novels. It was the classic case of a character in search of an author. I, however, did not see things quite that clearly because Zorro isn't like any of my protagonists; he wasn't a subject I would have chosen. With the last book of the trilogy, I had considered the experiment with juvenile novels ended. I had discovered that I prefer to write for adults; it has fewer limitations. A book for a young reader requires as much work as one for adults but you have to be prudent in matters referring to sex, violence, evil, politics, and other elements that give flavor to a story but that the editors do not think appropriate for that age. It irks me to have to write with "a positive message." I don't see any reason to protect young people, who already have a lot of filth in their heads; on the Internet they can see fat women fornicating with burros, or narcotraffickers and police torturing each other with the greatest ferocity. It is disingenuous to stuff positive messages into the pages of a book: the only thing that will be achieved is turning off all the readers. Zorro is a positive character, the hero par excellence, a mixture of Che Guevara, obsessed with justice, Robin Hood, always ready to take from the rich to give to the poor, and Peter Pan, forever young. I would have to work really hard to make a villain out of him, but, as his owners explained, that wasn't what it was about. Further, they warned me that the novel could not contain explicit sex. To put it briefly, it was a great challenge. I thought it over very carefully and in the end resolved my doubts in the usual way: I tossed a coin in the air. And that was how I ended up in my study with Diego de la Vega for several months.

Zorro had been greatly exploited; there wasn't much left to tell except his youth and old age. I opted for the former, since no one likes to see his hero in a wheelchair. What was Diego de la Vega like as a boy? Why did he become Zorro? I researched the historical period, the early 1800s, an extraordinary time in the Western world. The democratic ideals of the French Revolution were transforming Europe, and from them were born the wars of independence of colonies across the Americas. The victorious armies of Napoleon invaded several countries, including Spain, where the population rebelled and waged a bloody war that finally drove the French from Spanish soil. It was the era of pirates, secret societies, slave traffic, Gypsies, and pilgrims. In California, by contrast, nothing novelistic was happening; it was a vast rural expanse with cows, Indians, bears, and a few Spanish colonists. I would have to take Diego de la Vega to Europe

My research turned up more than enough material, and the protagonist already existed. My task was to create the adventure. To that purpose, Willie and I visited New Orleans, to follow the trail of the celebrated pirate Jean Lafitte, and we were lucky to know that exuberant city before Hurricane Katrina reduced it to a national shame. In the French Quarter, night and day, we heard brass bands and banjos, the golden voices of the blues, the irresistible call of jazz. People drank and danced to the hot rhythm of the drums in the middle of the street: color, music, the smell of food, and magic. There was enough for a whole novel, but I had to limit Zorro to a brief visit. I try now to remember New Orleans as it was then, with its pagan carnival in which people of every sort blended together in dance; New Orleans, with its venerable residential streets and centuries-old trees—cypresses, elms, magnolias in bloom—the wrought-iron balconies where two hundred years ago the most beautiful women in the world sat to enjoy the cool evening air, the granddaughters of Senegalese queens and the masters of the time, sugar and cotton barons. But the most lasting images of New Orleans are those of the recent hurricane: floods of filthy water and its citizens, always the poorest, struggling against the devastation of nature and the negligence of authorities. They became refugees in their own country, abandoned to their fate while the rest of the nation, stunned by scenes that seem as remote as a monsoon in Bangladesh, wondered whether governmental indifference would have been the same had the injured parties been white.

I fell in love with Zorro. Although I could not recount his erotic feats in the book in the detail I would have liked to, I can imagine them. My favorite sexual fantasy is for that appealing hero to quietly climb to my balcony, make love to me in shadows with the wisdom and patience of Don Juan, undeterred by my cellulite and my years, and disappear at dawn. I lie drowsing between the wrinkled sheets without a hint of who the gallant was who so favored me, because he never took off his mask. No guilt assigned.

Summer

Summer came with its usual pandemonium of bees and squirrels. The garden was at its peak, as were Willie's allergies: he will never give up counting the petals of each rose. But that has never stood in the way of his monumental achievements at the grill, something Lori also participates in; she left behind her many years as a vegetarian when Dr. Miki Shima, as much a vegetarian as she, convinced her that she needed more protein. Our heated swimming pool attracted hordes of children and visitors. The days grew longer under the sun, slow, with no clock, like days in the Caribbean. Tabra was the only one who was missing; she was in Bali, where they make some of the pieces she uses in her jewelry. Lagarto-Emplumado went with her for a week, but he had to return to California because of his fear of snakes and the packs of hungry, mangy dogs. It seems that he was opening the door to his room and a little green snake slithered by, brushing his hand. It was one of the most lethal snakes there are. That same night something warm, moist, and furry dropped from the roof, landed on them, and then scurried out of the room. They couldn't turn on the light in time to see it. Tabra said that it must have been an opossum, and she punched her pillow a few times and went back to sleep, while Lagarto spent the rest of the night on guard, with all the lights on and his knife in his hand, with no idea what an opossum might be.

Juliette and her boys spent several weeks with us. Aristotelis is the most polite and considerate member of the tribe. He was born with a slight tendency toward tragedy, like any Greek worth his salt, and from the time he was a boy had taken on the role of being his mother's and his brother's protector. Contact with the other children had lightened his load, and he became a comedian. I think he will be an actor, for in addition to being dramatic and handsome, he gets all the leads in school plays. Achilleas is still a little angel, prodigal with smiles and kisses; we spoil him outrageously. He swims like an eel and can spend twelve hours in the water. We pull him out wrinkled and sunburned and make him go pee in the bathroom. I don't like to think what all must be in that water. "Don't worry, ma'am," I was reassured by the pool maintenance man when I shared my doubts with him. "The chlorine content is so high that you could have a corpse in there and still have no problem."

The kids changed day by day. Willie had always said that Andrea had the same features Alejandro has, but just a little askew, and that one day they would settle into place. Apparently that's what was happening, though she paid no attention because she lives in her own world, dreaming, with her nose in her books, lost in impossible adventures. Nicole has turned out to be very smart, an excellent student, as well as sociable, friendly, and a flirt, the only one in the matriarchal tribe with that quality; none of the rest of us is dying to seduce anyone. With her esthetic instinct, she can with one appraising look destroy any woman's pleasure in what she's wearing—with the exception of Andrea, who is indifferent to fashion and is always in some costume or other, her style since early childhood. For months we watched Nicole going about carrying a mysterious black case, and we prodded her so hard that finally one day she showed us what was inside. It was a violin, which she had borrowed at school because she wanted to join the orchestra. She placed it to her shoulder, took up the bow, closed her eyes, and left us awestruck with a short and impeccably performed concert of melodies we had never heard her practice. Alejandro's skeleton shot upward with a great spurt, just in time, because I was planning to have the doctor give him growth hormones, the way they do cows, so he wouldn't end up a shrimp. I was afraid that he was the only one of my descendants to inherit my undesirable genes, but that year, to our relief, he had saved himself. Although he already had the shadow of a mustache, he was still behaving like a madcap, making faces in mirrors and bothering everyone with inopportune jokes, determined to avoid at any cost the anguish of growing up and taking care of himself. He had announced that he planned to live with his parents, one foot in each house, until he was married or was kicked out. "Hurry and grow up before we lose patience," we often warned him, tired of his clownish pranks. The twins hit the pool in two plastic floating turtles, observed from afar by Olivia, who never lost hope they would drown. Of all the fears that dog had when she came into our family, only two remain: umbrellas and the twins. All these little ones and the friends who often came with them ended the summer as tan as Africans, their hair turned green from the chemicals in the pool that are so lethal they burned the grass. Anywhere the swimmers set their wet feet, no grass would grow.

My grandchildren were at an age to discover love, all of them except Achilleas, that is, who was still at the stage of asking his mother to marry him. The kids hid in the nooks and crannies of the House of the Spirits and played in the dark. Their conversations in the pool often made their parents uneasy.

"Don't you know that you've broken my heart?" Aristotelis asked, breathing heavily through his mask.

"I don't love Eric anymore. I can come back to you, if you like," Nicole proposed between dives.

"I don't know, I have to think about it. I can't go on suffering like this."

"Well, think quick, otherwise I'll call Peter."

"If you don't love me, I might as well just kill myself today!"

"Okay. But don't do it in the pool. Willie will have a fit."

Rites of Passage

That summer of 2005 I finished writing *Inés of My Soul* and sent the manuscript off to Carmen Balcells with a big sigh of relief because it had been such a big project, and then, with Nico, Lori, and the children, we went on safari to Kenya. For several weeks we camped among the Samburu and Masai to watch the migration of the wildebeests, millions of them with the look of black cows racing terror-stricken across the Serengeti to Masai-Mara, a time of orgy for other animals that come to feast on the laggards. In one week nearly a million calves are born in this stampede. From fragile little airplanes we saw the migration like a gigantic shadow spreading across the African plains below. Lori had conceived the plan to take the children every year to some unforgettable place that would pique their curiosity and demonstrate that despite distances people are alike everywhere. The similarities that unite us are much greater than the differences that separate us. The year before we had gone to the Galapagos Islands, where the children swam with sea lions, turtles, and manta rays; Nico swam for hours out in open water behind sharks and whales as Lori and I ran from pillar to post looking for a boat we could use to rescue him from a sure death. By the time we found one, Nico was already swimming back with strong, steady strokes. We had hauled to Kenya Willie's cases of photographic equipment, his tripods, and the gigantic lens that never once captured an African beast because it was too clumsy to handle. Nicole took the best picture of the trip with a disposable camera: the eighteen-inch, blue-tongued kiss a giraffe planted on my face. Willie's heavy lens was eventually left in the tent while he used other more modest ones to immortalize the always quick smile of the Africans; their dusty markets; the five-year-old children tending the family herd in the middle of nowhere at three hours' march from the nearest village; the lion cubs and slim giraffes. In an open jeep we drove among the herds of elephants and buffalo, to muddy rivers where whole families of hippopotamuses were disporting themselves, and followed the wildebeests in their inexplicable race.

One of our guides, Lidilia, a pleasant Samburu with snow-white teeth and three long feathers crowning the bead adornment he wore on his head, became Alejandro's friend. Lidilia proposed to Alejandro that he stay there with him and be circumcised by a tribal witch doctor as the first step in his rite of passage. After that he would have to spend a month alone in the wild, hunting with a lance. If he succeeded in killing a lion, he could choose the most desirable girl in the village, and his name would be recorded along with those of other great warriors. My grandson, terrified, counted the days till he could get back to California. We called on Lidilia to translate when a warrior of some years wanted to buy Andrea as a wife. He offered us several cows for her, and since we refused, he added some sheep. Nicole telepathically communicated with the guides and the animals, and she has a notable memory for details, so she kept us informed: elephants change a complete set of teeth every ten years until they are sixty, at which time no new ones come in and they are condemned to die of hunger; a male giraffe measures nineteen and a half feet in height, his heart weighs thirteen pounds, and he eats a hundred and thirty-two pounds of leaves every day; among antelopes, the alpha male must defend his harem from his rivals and mate with the females, which leaves him very little time to eat, so he grows weak and another male bests him in combat and drives him off. The position of alpha male lasts about ten days. By then Nicole knew what it was to mate. Even though I am not made for the rustic life, and nothing horrifies me so much as not having an available mirror, I couldn't complain about the comfort of the trip. The tents were deluxe, and thanks to Lori, who arranges things to the last detail, we had hot water bottles in our beds, miner's lamps so we could read on dark nights, lotions to fend off mosquitoes, an antidote for snakebite, and every afternoon we had English tea served in a porcelain teapot as we watched a pair of crocodiles devour a forsaken gazelle.

Back in California, before summer ended, Alejandro had his rite of passage, though it was somewhat different from the one the Samburu Lidilia had proposed. He signed on for a program Lori and Nico discovered on the Internet, and once the four parents were convinced that it was not a ploy of some pedophiles and sodomites, they allowed him to go. Just as Lidilia had explained, every male should go through a ceremony marking his passage from child to adult. In lieu of such a tradition, several instructors had organized a three-day retreat in the woods in which boys would reinforce the concepts of respect, honor, courage, responsibility, the obligation to protect the weak, and other basic norms that in our culture tend to be relegated to medieval chivalric novels. Alejandro was the youngest of the group. The night he left I had a terrifying dream: my grandson was sitting by a bonfire among a crew of hungry orphans shivering with cold—a scene from a Dickens tale. I implored Nico to go get his son before something awful happened in these sinister woods where he was camping with a gang of strangers, but he paid no attention to me. At the end of the three days he went to get him, and they got back in time for Sunday dinner at the family table. We'd cooked beans, using a Chilean recipe, and the house smelled of corn and sweet basil.

The whole family was waiting for the initiate, who arrived filthy and hungry. Alejandro, who for years had said that he didn't want to grow up, seemed older. I hugged him with a grandmother's frenetic love, told him my dream, and it turned out that his experience had not been exactly like that, although there were a few orphans and a bonfire. There were also some delinquents who, according to my grandson, "were good kids who had done some stupid things because they didn't have any family." He told us that they'd sat in a circle around the fire, and each one told what caused him pain. I proposed that we do the same thing, since we were in a tribal circle, and one by one we gave our answer to the question posed to Alejandro. Willie said that he was anguished over the situation of his children: Jennifer, lost to us, and his two sons on drugs; I spoke of missing you; Lori, of her infertility . . . and all of us in turn exposed our deepest sorrow.

"And what gives you pain, Alejandro?" I asked.

"My fights with Andrea. But I've decided to get along better with her, and I will, because I learned that we're responsible for our own sadness."

"That isn't always true. I'm not responsible for Paula's death, or Lori for her infertility," I argued.

"Sometimes we can't avoid sadness, but we can control our reaction to it. Willie has Jason. As for you, because you lost Paula you created a foundation and you've kept her memory alive among us. Lori couldn't have children, but she has the three of us," he said.

Forbidden Love

Juliette didn't work during those months she lent her body to carry Lori and Nico's baby because she had to subject herself to the scourge of fertility drugs. The family had taken on the responsibility of looking after her, but once that dream was put aside she went out to look for work. She was hired by a broker who was planning to buy Asian art in San Francisco for his galleries in Chicago. Ben was fifty-seven well-lived years old, and he must have had a lot of money because he was as splendid as a duke. He planned to commute frequently from Chicago, and in his absence have someone in California look after the precious objects imported from the East. At the end of his first interview with Juliette, he invited her to dinner at the best restaurant in Marin County, a yellow Victorian house set among pines and masses of climbing roses. After several glasses of white wine, he decided that not only was she the ideal assistant, he was taken with her personally. By a coincidence worthy of a novel, Juliette learned during their conversation that Ben had known Manoli's first wife, the Chilean who had run off with her yoga instructor on her wedding day. He told Juliette that the woman was living in Italy and was in a fourth marriage to a magnate in olive oil.

It had been an eternity since Juliette had felt desired. The year before he died, Manoli had gradually ceased to be the passionate lover who had seduced her when she was twenty; his illness was corroding his bones and his spirit. Ben proposed to fill that void, and we watched Juliette come to life, resplendent, with a new light in her eyes and a mischievous smile dancing on her lips. Her life was turned upside down; she was taken to expensive places, restaurants, walks, theater, opera. Ben showered gifts and attention on Aristotelis and Achilleas. He was such an expert lover that he could make her happy over the telephone; that made his absences bearable, and when he came to California she would be eagerly awaiting him. Lori and I used one of our quiet little mid-afternoon breaks, with jasmine tea and dates, to corral her. It seemed to us that her attitude was slightly evasive, but we didn't have to press too hard before she told us about her affair with her boss. I heard that alarm bell that comes with experience, and I put in my two cents to warn her that she shouldn't mix work with love, because she could lose both. "He is using you, Juliette. How convenient! He has an assistant and a lover for the price of one," I told her. But she was already trapped. Both Lori and I had noticed that Juliette attracted men who had nothing to offer her: they were married, too old for her, too far away, or incapable of making a commitment. Ben might be one of those, because to us he seemed a little slippery. According to Willie, in today's hedonistic California, no man would take on the responsibility of a young widow with two small sons, but according to my astrologer, whom I had consulted in secret so I wouldn't be laughed at, it was all a matter of waiting three or four years, when the planets would send Juliette the ideal husband. Ben had moved in ahead of the planets.

When we returned from Kenya, Juliette's amorous fling had become more complicated. It turned out that Ben had not earned his fortune with a good eye for art; his wife had inherited it. The galleries were just a diversion to keep him occupied and riding the crest of the social wave. Ben's frequent trips to San Francisco and his whispered telephone conversations were beginning to raise his wife's suspicions.

"It isn't a good idea to get involved with a married man, Juliette," I told her, remembering the foolish things I had done when I was young, and the price I had paid.

"It isn't what you're thinking, Isabel. It was inevitable; we fell in love at first sight. He didn't seduce me or deceive me, it all happened by mutual consent."

"What are you going to do now?"

"Ben has been married for thirty years. He greatly respects his wife and adores his children. This is his first infidelity."

"I have the feeling that he's a chronic adulterer, Juliette, but that isn't your problem, it's his wife's. Yours is to look out for yourself and your sons."

To convince me of her gallant's honesty and his feelings for her, Juliette showed me his letters, which to me seemed suspiciously prudent. They weren't love letters, they were legal documents.

"He's covering his back. Maybe he's afraid you will charge him with sexually harassing you on the job, and that's illegal here. Anyone who reads these letters, including his wife, would think that you took the initiative, that you trapped him, and that you're the one doing the chasing."

"How can you say that!" she exclaimed, shocked. "He's waiting for the opportune moment to tell his wife."

"I don't think he'll do that, Juliette. They have children and a lifetime together. I feel sorry for you, but I feel sorrier for his wife. Put yourself in her place; she's a middle-aged woman with an unfaithful husband."

"If Ben isn't happy with her—"

"He can't have everything, Juliette. It's up to Ben to choose between you and the good life she offers him."

"I don't want to be the cause of a divorce. I told him that he should try and reconcile with his wife, that they should both go to therapy, or that he should take her to Europe on a second honeymoon," she said and burst into tears

I was afraid that at that rate the game would go on until the thread broke at the weakest point: Juliette. I didn't insist any further, though, because I was afraid I would drive her away. Besides, I am not infallible, as Willie reminded me, and it might be that Ben was really in love with her and that he would get a divorce in order to be with her, in which case, I, for behaving like a bird of bad omen, would lose the friend I had come to love like another daughter.

Just as we feared, Ben's wife traveled from Chicago to sniff the air in San Francisco. She installed herself in the office of her husband, who had the good sense to disappear, using a variety of excuses, and within a few hours, her instinct, and what she knew about him, had confirmed her worst fears. She decided that her rival could be no other than his beautiful assistant, and she confronted Juliette with all the weight of her authority as Ben's legitimate wife, along with the confidence lent by money and her pain, which Juliette could not ignore. The wife fired her, and warned her if she ever tried to communicate with Ben, she herself would see that something bad happened to her. Ben hadn't shown his face; he limited himself to a phone call, offering Juliette a small compensation and asking her, if you can believe it, to train her replacement before she left. His wife supervised that call, and the whining letter, last of the series, that closed the episode.

Two days later Willie came home to find Lori and me in the bathroom holding Juliette, who was curled up on the floor like a whipped child. We brought him up-to-date on what had happened. He said that he had seen it coming; it was not an original drama and everyone recovers from a broken heart. Within a year, he said, we would all be enjoying a glass of wine and having a good laugh as we recalled this unfortunate adventure. However, when Juliette told Willie of the wife's threats, he found it less amusing. He offered to represent her; she was, after all, entitled to file suit. The case could not be more attractive to a lawyer: a young widow with no money, mother of two young boys, and the victim of a millionaire employer who sexually used and then dismissed her. Any jury would crucify Ben. Willie already had the knife between his teeth, but Juliette wouldn't hear of that possibility because it wasn't true. They had fallen in love, she wasn't a victim. She did allow Willie to send a scorching hot letter announcing that if they threatened Juliette again, the matter would be taken to court. On his own initiative, Willie added that if the wife wished to resolve the problem, she should control her husband better. A letter wouldn't stop her if she was the kind of person capable of hiring mafiosi to do harm to a rival, but it proved that Juliette was not without protection. In less than a week, a lawyer in Chicago called to assure Willie that there had been a misunderstanding and there would be no further threats.

Juliette suffered for months, wrapped in the tight embrace of the family, but I wouldn't be recounting this lamentable episode had she not given me permission to do so, and had Willie's prognostication not come to pass. I hired Juliette to be my assistant; she began studying Spanish and soon was a participant in the Sausalito literary brothel, where she could work in peace with Lori, Willie, and Tong, who charged themselves with protecting her and with keeping at bay any unfaithful husband who rang the doorbell with lustful intentions. Before the year ended, one night when the whole family was eating dinner at the table of the Mistress of the Castle, Juliette lifted her glass in a toast to love affairs of the past. "To Ben!" we said in a single voice, and her laugh was loud and heartfelt. Now I am waiting for the alignment of the planets that will bring the good man who will make this young woman happy. It's likely that that could be happening soon.

Abuela Hilda Leaves
with You

For some time, Abuela Hilda had lived with her daughter in Madrid, where she and her second husband were carrying out a diplomatic mission. It had been a year since this peerless grandmother had come for one of her long visits with us; she had aged suddenly and was afraid to travel alone. In the 1960s, in Chile, I was a young journalist juggling three jobs at once to survive, but the births of my two children didn't complicate my life since I had help. In the mornings, before going to work, I went by and left you either at the house of my mother-in-law, our adorable Granny, or with Abuela Hilda, who took you, still asleep and bundled in a shawl, and looked after you all day until I came to pick you up in the evening. When you started school, it was Nico's turn; he too was cared for by those fairy-tale grandmothers who spoiled him like the firstborn of an emir. Following the military coup, we went to Venezuela, and what you and Nico missed most were those two grandmothers. Granny, whose only life was her grandchildren, died of sorrow a couple of years later. When Abuela Hilda was widowed, she came to Venezuela since her only daughter, Hildita, lived there, and took turns between Hildita's house and ours. My relationship with Abuela Hilda had begun when I was seventeen. Hildita was my brother Pancho's first girlfriend; they met at school when they were fourteen, ran away, were married, had a son, divorced, married again, had a daughter, and were divorced a second time. In all, they spent more than a decade loving and hating each other, while Abuela Hilda witnessed the spectacle without comment. I never heard her say a disapproving word against my brother, who perhaps deserved it.

At some moment of her life, Abuela Hilda had decided that her role was to help her small family, in which she generously included me and my children, and she did it to perfection, thanks to her proverbial discretion and good health. She was as strong as a mule, which was why she was able, for example, to take you, Nico, and another half dozen teenagers camping on a Caribbean island that had no water. They reached it by crossing a treacherous sea in a small boat, followed closely by sharks. The boatman left them with a mountain of equipment and, to their good luck, remembered to pick them up a week or two later. Abuela Hilda survived the mosquitoes, the nights drinking rum and warm Coca-Cola, the canned beans, the aggressive mice that nested in their sleeping bags, and other inconveniences that I, twenty years younger, could never have endured. With the same magnanimous attitude, she sat herself down to watch pornography when you were studying psychology and decided to specialize in sexuality. You went around everywhere carrying a suitcase filled with paraphernalia for erotic games that seemed in very bad taste to me, but I never dared voice my opinion, fearing that you would have teased me unmercifully for being such a prude. Abuela Hilda sat down with you, knitting without looking at her needles, and watched some hair-raising videos that included trained dogs. She was an active member of our ambitious home theater company; she sewed costumes, painted scenery, and played any role she was asked, from Madama Butterfly to Joseph in our Christmas pageants. As time passed she grew smaller and her voice thinned to a birdsong, but her enthusiasm for participating in family madness never faltered.

Abuela Hilda's last days were not spent with us but with her daughter, who cared for her during her rapid decline. It began with repeated bouts of pneumonia, a vulnerability left from her days as a smoker, the doctors said, and after that she began forgetting her life. Hildita recognized her mother's final stage as a return to childhood and decided that if you can squander patience on a two-year-old child, there was no reason to deprive an old woman of eighty of the same indulgence. She watched lovingly to see that her mother bathed, ate, took her vitamins, and went to bed. She had to answer the same questions ten times and pretend that she was listening when her mother finished telling a meaningless anecdote and then like a recording repeated the same words over and over. Finally, Abuela Hilda tired of swimming through a nebula of confused memories, of being afraid to be alone, of falling, of the creaking of her bones, and of the assault of faces and voices she couldn't identify. One day she stopped eating. Hildita called me from Spain to tell me what a battle it was to feed her mother a spoonful of yogurt and the only thing I could think to tell her was not to force her. That was how my grandfather died; he lost his appetite when he decided that one hundred years was too much living.

Nico caught a plane to Madrid the day after Hildita called. Abuela Hilda knew who he was immediately, even though she didn't recognize herself in the mirror.

Feeling suddenly flirtatious, she asked for her lipstick, and suggested they play a game of cards, which was accomplished with their usual cheating. Nico got her to drink a warm Coca-Cola with rum, in homage to the Caribbean adventures, and in the next half hour he fed her a small bowl of soup. The visit of her proxy grandson, and the promise that if she gained weight she could come to California and smoke marijuana with Tabra, worked the miracle. Abuela Hilda began to eat again, but that lasted only a month or two more. When she again declared a hunger strike, her daughter decided with sorrow that her mother had every right to go in her own way and at her own time. Abuela Hilda, who was always a small, slim woman, in the next weeks became a minuscule, big-eared sprite so light she could be lifted up on the breeze through the window. Her last words were, "Hand me my purse; Paula came to get me and I don't want to make her wait."

I reached Madrid a few hours later, but not in time to help her daughter take care of the details demanded by death. A few days later I returned to California with a small box containing a handful of Abuela Hilda's ashes to scatter in your forest; she wanted to be with you.

Reflections

I began these pages in 2006. My January 8 ritual has become more complicated with the years; I no longer have the arrogant certainty of youth. To throw myself into another book is as grave as falling in love, a crazed impulse that demands fanatic dedication. With each one, as with a new love, I wonder whether I will have the strength to write it, even whether the project is worth the trouble; there are too many pointless pages, too many frustrated affairs. In the past I submersed myself in writing—and in love—with the temerity of someone who ignores the risks, but now it takes several weeks before I lose my respect for the blank screen of the computer. What kind of book will this be? Will I make it to the end? I don't ask myself those questions about love because I've been with the same lover for eighteen years and have banished any doubts; now I love Willie every day without questioning what kind of love it is or how it will end. I want to believe that it's an elegant love, and that it will not have a vulgar ending. Maybe what Willie says is true, that we will go hand in hand to the other side of death. All I want is for neither of us to lose our way in senility and cause one of the partners to care for a decrepit body. To live together, lucid, to the last day; that would be ideal.

The ritual of beginning another book is more or less the same every year. So I thoroughly cleaned my study, aired it out, changed the candles on what my grand-children call "the ancestor altar," and got rid of boxes filled with texts and documents used in researching last year's undertaking. I left nothing on the shelves lining the walls other than the tightly aligned first editions of my books and pictures of the living and dead who are always with me. I took out anything that might muddle inspiration or distract me from this memoir that demands clear space in which to express itself. It was the beginning of a time of solitude and silence. I always take a while to get started; at first the writing moves along in ragged spurts, like a rusty machine, and I know that several weeks must go by before the story begins to take shape. Any distraction frightens off the muse of imagination. What does imagination feed on, anyway?

In my experience, on memories, the vast world, the people I know, and also the persons and voices I carry within that help me on the journey of living and writing. My grandmother used to say that space is filled with presences, of what has been, is, and will be. My characters live in that transparent atmosphere, but I can hear them only if I am silent. Toward the middle of the book, when I am no longer me—the woman—but another—the narrator—I can see them as well. They emerge from the shadows and appear before me whole, with their voices and their smell; they assault me in my *cuchitril*, invade my dreams, occupy my days, and even follow me down the street. That doesn't happen with a memoir in which the protagonists are my own living family, filled with opinions and conflicts. In this case, the plot is not an exercise of imagination but an attempt to present the truth.

There was a sense of frustration in the country that had dragged on for a long time. The future of the world looked as dark and impenetrable as tar. The escalation of violence in the Middle East was terrifying, and international condemnation of America was unanimous, but President Bush paid no attention; he wandered like a madman, detached from reality and surrounded by sycophants. He could no longer obscure the calamity of the war in Iraq, even though the press showed only aseptic images of what was happening: tanks, green lights on the horizon, soldiers running through deserted villages, and occasionally an explosion in a market where supposedly the victims were Iraqis. No blood, no dismembered children. Correspondents were embedded in units of the troops and information was filtered through a military apparatus; however, on the Internet anyone who wanted to be informed could consult the media of the rest of the world, including Arab television. Some courageous reporters—and all the comedians and cartoonists—denounced the government's incompetence. Images of the prison at Abu Ghraib flew round the world, and in Guantanamo prisoners indefinitely detained without being charged died mysteriously, committed suicide, or agonized in hunger strikes, force-fed through large stomach tubes. Things were happening that could not have been imagined a short time before in the United States, which thinks of itself as a beacon of democracy and justice: the writ of habeas corpus was suspended for prisoners, and torture was legalized. I expected the public to react with one voice, but almost no one gave those matters the importance they deserved. I come from Chile, where for sixteen years torture was institutionalized: I know the irreparable harm that leaves in the souls of victims and victimizers—indeed the entire population, which becomes an accomplice. According to Willie, the United States had not been this divided since Vietnam. Republicans controlled everything, and if the Democrats didn't win in the November elections, we'd be screwed for good. How could they *not* win? I asked myself, considering that Bush's popularity had plummeted to numbers Nixon had in his worst days.

The person who was most disturbed was Tabra. When she was young she had left the country because she could not support the war in Vietnam, and now she was prepared to do the same thing, even to renounce her U.S. citizenship. Her dream was to end her days in Costa Rica, but a lot of foreigners had had the same idea and the price of property in that country had soared beyond Tabra's resources. That was when she decided to move to Bali, where she could conduct her business dealings with the local silversmiths and artisans. She would leave a couple of sales representatives in the United States and all the rest could be done over the Internet. That was all we talked about on our walks. Tabra saw fatal signs on every side, from the television news to mercury in salmon.

"Do you think it will be different in Costa Rica or Bali?" I asked her. "Wherever you go the salmon will have mercury, Tabra."

"At least there I won't be an accomplice to the crimes of this country. You left Chile because you didn't want to live under a dictatorship. Why can't you understand that I don't want to live here?"

"This isn't a dictatorship."

"But it can become one, sooner than you think. What your Tío Ramón told me is true: people get the government they deserve. That's the downside of a democracy. You should leave, too, before it's too late."

"My family is here. I've put a lot into bringing them together, Tabra, and I want to enjoy them because I know it can't last much longer. Life tends to separate us, and it takes a lot of effort to stay together. At any rate, I don't think we've reached the point where it's necessary to leave this country. We can still change things. Bush won't be around forever."

"Well, good luck. As for me, I'm going to settle down in some peaceful place where you can come with your family when you need somewhere to go."

I began gradually to tell Tabra good-bye as she dismantled the workshop it had cost her so many years to establish. She had help from her son, Tangi, who left his job to be with her in her last months in this country. One by one she said good-bye to the refugees she had worked with for so long, worried about them because she knew that for some it would be difficult to find another job. She got rid of her art collections, with the exception of some valuable paintings I'm keeping for her. She couldn't completely cut her ties with the United States. She would have to come back at least a couple of times a year to see her son and to supervise her business interests; her jewelry requires a much larger market than tourist beaches in an Asian paradise. I assured her that when she came to California she could always count on having a room in our home. Then she emptied her house of furniture and put it up for sale.

Those preparations and my sad walks with Tabra infected me with the delirium of my friend's uncertainty. I would go home and hug Willie, feeling blue. Maybe it wasn't such a bad idea to put our savings into gold coins; we could sew them into the hem of a skirt and be ready to flee. "What gold coins are you talking about?" Willie asked me.

The Tribe Reunited

Andrea's entrance into adolescence was sudden and dramatic. One night in November she came into the kitchen, where the family was gathered, wearing contact lenses, lipstick, a long white dress, silver sandals, and drop earrings made by Tabra; she had been chosen to sing in the chorus at the school Christmas festivities. We didn't recognize that sensual, golden beauty from Ipanema with her distant, mysterious air. We were used to seeing her in scroungy blue jeans and clumsy outback boots, with a book in her hand. We'd never seen that girl who was shyly smiling at us from the doorway. When Nico, whose Zen serenity we had so often laughed at, realized who it was, he was thunderstruck. Instead of celebrating the young woman who'd just made an appearance, we had to console her father over the loss of his awkward little girl. Lori, who had taken Andrea to buy the dress and the makeup, was the only one in on the secret of the transformation. While the rest of us were recovering from our stupefaction, she took a series of photographs, some with Andrea's thick dark honey-colored hair loose on her shoulders, some with it piled on her head, in a model's poses that were all affectation and spoof.

Andrea's eyes were shining, and she was flushed, as if she'd been in the sun, though the rest of us were wearing our November pallor. She'd had a bad cough for several days. Nico wanted to have a picture with her sitting on his knees, the same pose as one when she was five and she was a plucked duck wearing an alchemist's thick eyeglasses and the pink nightgown she wore over her normal clothes. When he touched her, she was burning hot. Lori took her temperature and the small family party turned dark, because Andrea was aflame with fever. Within a few hours she was delirious. They tried to bring down her fever with cold baths, but finally they rushed her to the emergency room, where they learned she had pneumonia. Who knows how many days she had been incubating it and hadn't said a word, faithful to her stoic, introverted nature. "My chest hurts, but I thought it was because I'm developing," was her explanation.

Celia and Sally came immediately, then the others. Andrea was in the local hospital surrounded by family, all of us watching like hawks to be sure that she wasn't given anything on the porphyria blacklist. Seeing her in that iron bed, her eyes closed, her eyelids transparent, growing paler every moment, breathing with difficulty, and connected to tubes and wires, brought back my worst memories of your illness in Madrid. Like Andrea, you checked into the hospital with a bad cold, and when you left six months later, you were no longer yourself but a lifeless doll whose only hope was for a gentle death. Nico calmly reasoned with me: this wasn't the same. You had terrible stomach pains for several days, and couldn't eat without vomiting, porphyria crisis symptoms that Andrea did not have. We decided that to avoid any possible oversight or medical error, Andrea would never be alone. We hadn't been able to do that in Madrid, where the hospital bureaucracy had taken you over with no explanation. Your husband and I stood guard for months in a corridor, never knowing what was happening on the other side of the heavy doors of the ICU.

Andrea's room in the hospital was filled. Nico and Lori, Celia and Sally, and I installed ourselves at her side. Then Juliette came, Sabrina's mothers, the other relatives, and a few friends. Fifteen cell phones kept us connected, and every day I called my parents and Pía in Chile so they would be with us, though far away. Nico handed out the list of forbidden medications and instructions for each eventuality. Your gift, Paula, was that now we were prepared; we wouldn't let anything take us by surprise. Our doctor, Cheri Forrester, asked the personnel on the floor to be forbearing because this patient came with a tribe. While the nurse was pricking Andrea, looking for a vein to place an IV, eleven people around the bed were watching. "Please, just don't chant," said the nurse. We all laughed. "You look like the kind of people capable of chanting," she added, preoccupied.

The day-and-night vigil began, never fewer than two or three of us in the room. Very few went to work; those who weren't taking a turn at the hospital were looking after the other children and the dogs—Poncho, Mack, and especially Olivia, who was a nervous wreck from finding herself shunted aside—keeping the houses running, and bringing food to the hospital to feed our army. For two weeks, Lori assumed the role of captain, which no one tried to usurp because she actually is the manager of this family and I don't know what we would do without her. No one has more influence or more dedication than Lori. Raised in New York, she is the only one with the intrepid character that will not be intimidated by physicians and nurses, that can fill out ten-page forms and demand explanations. In the last few years, we have moved past the obstacles of the first years; Lori is my true daughter, my confidante, my right arm in the Foundation, and I have watched how little by little she is being converted into the matriarch. Soon it will be her turn to take her place at the head of the table as the mistress of the castle.

Andrea at first grew weaker because they couldn't administer several of the antibiotics ordinarily used in such cases, and that prolonged the pneumonia longer than normal, but Dr. Forrester, ever vigilant, assured us there was no indication of porphyria in the blood and urine tests. Andrea would perk up for brief periods, when her brother and sister, the Greek boys, or some classmate visited, but the rest of the time she slept and coughed, holding the hand of one of her parents or her grandmother. Finally, the second Friday of her stay, her temperature was normal and she woke up with clear eyes and an appetite. We could draw a breath of relief.

For more than ten years the family had been dancing its way through the skirmishes that tend to follow divorces, a tug-of-war that leaves everyone exhausted. The relationship between the two sets of parents had its ups and downs; it was not easy to come to agreements on details of raising the children they were sharing, but as they were moving out into the world to make their own lives, there are fewer reasons for confrontation, and the day will come when there is no further need to see each other. That day is fairly close. Despite the difficulties they've had, they can congratulate one other because they have three happy and likable children who get good grades and are well behaved, and who up to this moment have not presented a single serious problem. During the two weeks of Andrea's pneumonia, I lived the dream of a united family; it seemed to me that all the tensions had disappeared beside our little girl's bed. But in real-life stories there are no perfect endings. Each person just has to do the best he or she can, that's all.

Andrea left the hospital weighing ten pounds less, listless and the color of a cucumber, but more or less recovered. She spent another two weeks convalescing at home, and got well in time to sing in the chorus. From the audience we watched her come in singing like an angel in the long row of girls who occupied the stage. The white dress hung on her like a rag and her sandals fell off her feet, but we were all in agreement that she had never been prettier. The entire tribe was there to celebrate her, and once more I found that in an emergency you toss overboard the things that are not essential, that is, nearly everything. In the end, after a thorough lightening of loads and taking account, it turns out that the one thing that's left is love.

A Time to Rest

We have come to December and the panorama has changed for our tribe and for the country. Tabra went off to Bali; my parents, in Chile, are living on borrowed years; they are eighty-five and ninety; Nico finally turned forty—at last, as Lori says—and is a mature man; the grandchildren are full-blown teenagers and soon will be weaning themselves from the obsessive grandmother who still calls them "my babies." Olivia has grown gray hair and thinks twice about climbing the hill when we take her out for a walk. Willie is finishing his second book, and I am still plowing the hard ground of recollections in order to write this memoir.

The Democrats won the elections and now control both House and Senate; we all hope that will put the brakes on Bush's excesses, that the American troops will be removed from Iraq, even if gradually and with their tails between their legs, and that new wars can be prevented. As for Chile, there are changes there as well. In March, Michelle Bachelet assumed the presidency, the first woman in my country to occupy that post, and she is doing very well. She is a pediatrician, a socialist, a single mother, an agnostic, and the daughter of a general who suffered torture and death because he would not submit to the military coup in 1973. In addition, General Augusto Pinochet died tranquilly in his bed, thereby closing one of the most tragic chapters of our national history. With a great sense of timing, he died precisely on Human Rights Day.

Writing this book has been a strange experience. I have not relied exclusively on my memories and the correspondence with my mother, but have also questioned the members of my family. Since I write in Spanish, half of my family could not read it until it was translated by Margaret Sayers Peden—Petch—an adorable lady of eighty who lives in Missouri and has translated all my books except the first. With the patience of an archaeologist, Petch burrowed into the different layers of the manuscripts, reviewed each line a thousand times, and made the changes I asked. Once the manuscript was in English, the family could compare their different versions, which did not always coincide with mine. Harleigh, Willie's younger son, decided that he would prefer not to be in the book, and I had to rewrite it. That's a shame, really, because he is quite colorful and he's a part of this tribe; to exclude him seems to me to be cheating somehow, but I have no right to appropriate someone's life without his permission. In long conversations we were able to conquer our fear of expressing what we were feeling, the bad as well as the good; sometimes it is more difficult to show affection than anger. Which is the truth? As Willie says, you reach a point when you have to forget truth and concentrate on facts. As the narrator, I say that you have to forget facts and concentrate on the truth. Now that I'm coming to the end, I hope that this exercise of setting memories in order will be beneficial for everyone. Gently, the waters will settle, the mud will sink to the bottom, and there will be transparency.

Willie's and my lives have improved since the times of the marathon therapy sessions, the magical incantations to pay the bills, and the mission of saving from themselves people who didn't want to be saved. For the moment, the horizon is bright. Unless some cataclysm occurs, a possibility that can't be discounted, we are free to enjoy our remaining years lolling in the sun.

"I think we're old enough to retire," I said to Willie one night.

"No way. I've just begun to write and I don't know what we would do with you if you weren't writing. No one could put up with you."

"I'm serious. I've been working for a century. I need a sabbatical."

"What we *will* do is take things more calmly," he concluded.

Frightened by the threat of seeing me idle during a hypothetical sabbatical, Willie opted to take me on a vacation in the desert. He thought that a week in a sterile landscape with nothing to do would be enough to make me change my mind. The hotel, which according to the travel agency was four-star, turned out to be a kind of passé bordello where Toulouse-Lautrec would have felt right at home. We had reached the place by driving down an interminable expressway, a straight ribbon through naked desert dotted by improbably green golf courses baking beneath a white, incandescent sun still blazing at eight o'clock at night. There was no breeze, nor a bird in the sky. Every drop of water was transported from a great distance, and every blade of grass grew thanks to the inordinate labors of humble Latino gardeners who kept the complex machinery of this illusory paradise running and then disappeared at night like ghosts.

Fortunately, Willie suffered a near lethal allergy attack caused by the dusty drapes in that Toulouse-Lautrec bordello, and we had to go somewhere else. That's how we ended up at some strange hot springs we'd never heard of, where among other amenities they offered mud baths. Deep iron tubs were filled with a bubbling and boiling thick, fetid substance. A young Mexican Indian girl, short, square, with hair scorched by a cheap permanent, showed us around the installations. She wasn't more than twenty, but we were surprised by her self-confidence.

"What does this do?" I asked in Spanish, pointing to the mud.

"I don't know. It's something the Americans like."

"It looks like shit."

"It *is* caca, but not human; it's animal caca," she told me without a trace of irony.

The girl never took her eyes off Willie, and when we were about to leave, she asked him if he was Señor Gordon, the San Francisco lawyer.

"Don't you remember me, *licenciado?* I'm Magdalena Pacheco."

"Magdalena? But how you've changed, *niña!*"

"It's the permanent," she said, blushing.

They hugged each other, euphoric. She was the only daughter of Jovito Pacheco, Willie's client who had died in a construction accident years before. We went with her that night to have dinner at a Mexican restaurant where her older brother, Socorro, was king of the kitchen. He was married and already had a first son, a three-month old baby he'd named Jovito, after the child's grandfather. The other brother was working farther north in the Napa Valley vineyards. Magdalena had a Salvadoran boyfriend, a car mechanic, and said that she would be setting the date for their wedding as soon as the family could all meet in their village in Mexico; she had promised her mother she would be married in white in the presence of all their kin. Willie said that we'd be there, too, if they invited us.

The Pachecos told us that a year or two before, their grandmother was found dead one morning and they had given her an epic funeral, with a mahogany coffin the grandchildren had brought from San Diego in a truck. Apparently crossing the border in either direction wasn't a problem for them, even with a heavy casket. Their mother had a little store and lived with the youngest brother, the one who was blind, who by now was fourteen. On the way to the restaurant, Willie reminded me of the Pacheco case, which had dragged for years through the San Francisco court system. I hadn't forgotten, because we often teased Willie about the high-flown phrase he'd used in his argument: "Are you going to allow this defense lawyer to toss this poor family onto the garbage dump of history?" Willie had appealed from one judge to the next, until at last he won a modest settlement for the family. He had seen small fortunes disappear; clients who had never had anything but holes in their pockets often lost their heads when they felt they were rich. They had thrown money around and attracted distant relatives and forgotten friends like flies, and a bevy of swindlers eager to relieve them of their last peso. The Pachecos' award was a long way from being a fortune, but translated into Mexican pesos it had been a big help to the family, and had pulled them up from poverty. At Willie's suggestion, the grandmother had invested half in a small store and had deposited the rest in the name of Jovito's children in the United States, out of reach of bamboozlers and relatives with their hands out. It had been more than a decade since their father's death, and during that time all the children, except the youngest, had told their mother and grandmother good-bye and left their little village to work in California. Each came carrying a piece of paper with Willie's name and telephone number to claim the share of the money that belonged to them and would allow them to start a life under better conditions than most illegal immigrants, who came with nothing more than hunger and dreams. So Willie's stratagem when he took them to Disneyland as children had worked.

Thanks to Socorro and Magdalena Pacheco, we were given the best cabaña at the spa, an impeccable adobe and red-tiled casita in purest Mexican style; it had a small kitchen, a rear patio, and an open-air Jacuzzi. After buying provisions for three days, we shut ourselves in. It had been a long time since Willie and I had been alone, and idle, and we spent the first hours in invented tasks. With the minimal utensils in the kitchen, barely enough to put together a breakfast, Willie decided to cook oxtail, one of those slow Old World recipes that require several pots. His stew filled the air with a powerful aroma that frightened the birds and attracted the coyotes. Since it had to stay in the refrigerator until the next day so the fat that congealed on the surface could be spooned off, we dined, as night fell, on bread and cheese, lying close together in a hammock on the patio while the pack of coyotes licked their chops on the other side of the stone wall that protected our little haven.

A Quiet Place

Night in the desert has the unfathomable mystery of the bottom of the sea. An infinite embroidery of stars filled a black, moonless sky, and as the earth cooled, it emitted dense vapors like the breath of a wild beast. We lit three thick candles that spread their ceremonial light across the water of the Jacuzzi. Little by little the silence was relieving us of stress accumulated through so many scrapes and scraps. At my side there is always an invisible and implacable overseer, whip in hand, criticizing me and giving me orders: *Up, woman! Out of bed! It's six o'clock and you have to wash your hair and walk the dog. Don't eat that bread! Do you think you're going to lose weight by magic? You surely remember that your father was fat. You have to write your speech over, it's filled with clichés, and your novel is a disas-*

ter; you've been writing for a quarter of a century and haven't learned a thing. And on and on with the same tune. You used to tell me to learn to love myself a little, Paula, that I wouldn't treat my worst enemy the way I treat myself. "What would you do, Mamá, if someone came into your house and insulted you that way?" she would ask. I would tell him to go to hell and run him out with a broom, of course. I try to remember your advice, daughter, but it doesn't always work for me to use that tactic with the overseer because he's sly and he catches me off-guard. Luckily, on that occasion he had lagged behind in the little Toulouse-Lautrec hotel and wasn't there in our cabaña to bug me.

An hour went by, maybe two, in silence. I don't know what was going through Willie's mind and heart, but I was imagining there in the hammock that piece by piece I was shedding my rusted helmet, my heavy iron armor, my coat of mail, my leather breastplate, my studded boots, and all the pathetic weapons I'd used to defend myself and my family from the whims of fate. Ever since your death, Paula, I have often lost myself in your forest, taking tranquil walks on which you accompany me and invite me to search into my soul. In all these years it seems to me that the sealed caves inside me have been opening, and with your help light is falling in. Sometimes in that forest I sink into nostalgia and am invaded by a dull pain, but it doesn't last; soon I feel you walking beside me, and the rustling of the redwoods and the scent of rosemary and bay console me. I imagine how good it would be to die in this enchanted forest with Willie, old, but in full control of our lives and our deaths. Side by side, holding hands, here on this fragrant earth, we would abandon our bodies and join with the spirits. Maybe Jennifer and you will be waiting for us. If you looked for Abuela Hilda, I hope you won't forget to do the same for me. Those walks are very good for me, and at the end I feel invincible and grateful for the overwhelming abundance of my life: love, family, work, health—a great contentment. The experience of that night in the desert was different: I didn't feel the energy you give me in the forest but a letting go. Layers and layers of hard scales were sloughing off, and I was left with a vulnerable heart and weak bones.

About midnight, when the candles had nearly burned down, we took off our clothes and sank into the warm water of the Jacuzzi. Willie is no longer the same man who years before had attracted me at first sight. He still radiates strength, and his smile hasn't changed, but he is a man who has suffered; his skin is too white, his head shaved to disguise baldness, his eyes a paler blue. And on my face I carry the marks of past duels and losses. I had shrunk an inch and the body lolling in the water was that of a mature woman who had never been a beauty. But neither of us judged or compared; we didn't even look back to how we'd been in our youth. We have reached that stage of perfect invisibility that living together accords. We have slept together for so long that we no longer can see each other. Like two blind people, we touch, smell, sense the other's presence, the way you sense air.

Willie told me that I was his soul, that he had waited for me and looked for me the first fifty years of his life, sure that before he died he would find me. Willie is not a man to toss around pretty speeches; in fact he can be a little brusque, and he abhors sentimentality, and for that reason every one of his measured, carefully considered words fell over me like drops of rain. I realized that he, too, had entered that mysterious zone of the most secret surrender; he, too, had divested himself of his armor and, like me, opened his heart. I told him, in a thin voice, because he had taken my breath away, that without knowing it, I, too, had been feeling my way toward him. I have described romantic love in my novels, the love that gives everything, holding nothing back, because I always knew such love existed, though maybe it wasn't meant for me. The only taste I'd had of that total giving of self, that unconditional love, had been for you and your brother when you were very young. Only with you had I felt that we were a single spirit in barely separated bodies. Now I feel that with Willie. I have loved other men, as you know, but even in the most irrational passion I had guarded my back. From the time I was a little girl, I had looked after myself. In those games in the cellar of my grandparents' house where I grew up, I had never been the maiden rescued by the prince, only the Amazon who battled the dragon to save the town. But now, I told Willie, all I wanted was to lay my head on his shoulder and beg him to take care of me, as it seems men do with women when they love them.

"You don't think I take care of you?" Willie asked, startled.

"You do, Willie. You take care of all the practical things, but I'm talking about something more romantic. I don't even know exactly what. I guess I want to be the damsel in the fairy tale, and you to be the prince who saves me. I'm tired of slaying dragons."

"I've been your prince for almost twenty years, but you, my damsel, haven't noticed."

"That wasn't our agreement when we met; our deal was that I would look after myself."

"Did we say that?"

"Not in those words, but it was understood; we'd be comrades. But now the word *comrade* makes me think of guerrillas. I'd like to see how it feels to be your fragile wife for a change."

"Aha! Our Scandinavian instructor at the ballroom was right." He laughed. "The man leads."

My answer was to try to duck him; he pushed me and we both ended up under water. Willie knows me better than I know myself, and even so he loves me. We have each other, and that's something to celebrate.

"My God!" he exclaimed as he came up. "I was waiting in my corner, impatient because you didn't come, and you were waiting for me to invite you to dance! Is this why we had all that therapy?"

"Without the therapy I never would have admitted wanting you to look after me and protect me, my wanting to belong to you. How tacky! Think of it, Willie, this goes against a lifetime of feminism."

"It doesn't have anything to do with feminism. We need more private time, calm, time for just us. There's too much squabbling in our lives. Come with me to some quiet place," Willie murmured, pulling me to him.

"Some quiet place . . . I like that."

With my nose in his neck, I gave thanks for the good fortune of accidentally having found a love that so many years later has not lost its luster. Arms around each other, floating in the hot tub, bathed in the amber light of the candles, I felt that I was melting into this man with whom I had traveled a long, steep road, tripping, falling, getting up again, through fights and reconciliations, but never betraying each other. The sum of our days, our shared pains and joys, was now our destiny.

HARPER LUXE

THE NEW LUXURY IN READING

We hope you enjoyed reading
our new, comfortable print size and found it
an experience you would like to repeat.

Well – you're in luck!

HarperLuxe offers the finest in fiction and
nonfiction books in this same larger print size and
paperback format. Light and easy to read, HarperLuxe
paperbacks are for book lovers who want to see
what they are reading without the strain.

For a full listing of titles and
new releases to come, please visit our website:

www.HarperLuxe.com